A Conspiracy of Cells

A Conspiracy of Cells

One Woman's Immortal Legacy and the Medical Scandal It Caused

Michael Gold

State University of New York Press

Published by
State University of New York Press, Albany
© 1986 Michael Gold

For information, address State University of New York
Press, State University Plaza, Albany, N.Y., 12246

Library of Congress Cataloging in Publication Data

Library of Congress Cataloging-in-Publication Data

Gold, Michael, 1953—
    A conspiracy of cells.

    Bibliography: p.
    Includes index.
    1. Cancer — Research.   2. HeLa cells.   3. Cell
culture — Quality control.   4. Clone cells.   I. Title.
RC267.G64 1985   616.99'4'0072   85-26264
ISBN 0-88706-099-4
ISBN 0-88706-074-9 (pbk.)

10 9 8 7 6 5 4 3 2 1

To Susan West, my wife,
and to Martin and Helen Gold, my parents

Henrietta Lacks

If a man will begin with certainties,
he shall end in doubts;
But if he will be content to begin with doubts,
he shall end in certainties.

—Francis Bacon

Contents

1

Special Delivery

Jim Duff was carrying an odd little suitcase as he stepped out of the helicopter. It was a box of molded styrofoam, like a small ice chest you might take to a softball game to keep a six-pack cool. Except that this chest was made of thicker, more serious looking stuff.

Walter Nelson-Rees knew what was inside. It was the reason he had come out to the Berkeley heliport to meet Duff. The two men shook hands, and Nelson-Rees noticed that Duff was a bit edgy.

"We'd better go get a drink," said Duff.

When they had driven the few blocks to Oakland and found a bar, Duff gingerly placed the box onto the floor beneath their table and started his pitch. "Walter," he said in a quiet, earnest tone, "these cells are . . . different. And you're going to have to handle them . . . differently."

Of course the cells are different, thought Nelson-Rees. When had a bureaucrat from the National Cancer Institute ever flown from Washington, D.C., to California to personally deliver a chest full of dry ice and malignant tumor cells? There had been plenty of shipments over the years—hundreds, in fact—but air freight had always sufficed. The National Cancer Institute trusted the airlines, and when the airlines got the packages to Oakland, it trusted Nelson-Rees to handle the rest.

Tall, with a sharp nose and thin lips, Walter Anthony Nelson-Rees had a slightly aristocratic look about him and a manner that suggested he was sure of everything he did. He operated one of the best cell banks in the country, if not the best, and seemed quite aware of that. The officials of the National Cancer Institute knew it too. In fact, Nelson-Rees's cell bank existed largely to serve researchers working under the aegis of the institute. They were the bigshots, some of the best scientific minds in the country, and they needed the best weapons they could get to fight "The War."

"The War on Cancer," as the newspapers called it. The year was 1973. Richard Nixon wanted to be remembered as the president who brought cancer to its knees, and the National Cancer Institute, lavishly endowed, was his war department.

Yes, the institute wanted only the best for its boys, the choicest cuts of cancerous tissues from which they might extract the secrets of the disease, its cause and its cure. Nelson-Rees was the supply man. Cells from any organ taken from patients of any age, any race, either sex—you name it, he had it. When a researcher managed to get a new kind of cell growing in a culture dish, a cell that looked as though it could be useful for cancer experiments, chances are the institute would have a sample sent to Nelson-Rees.

He was as fastidious as *The Odd Couple*'s Felix Unger, which was the perfect recommendation for someone in charge of nurturing sometimes delicate cultures, keeping their identities straight, and protecting them from . . . well, from anything unexpected. Nelson-Rees complained regularly to maintenance workers at his lab about their inability to get the air pristine, free of dust and microorganisms. And few of those who saw it will ever forget the scene following Nelson-Rees's discovery

that a technician had used a hallway in the "clean section" of the cell bank to fold up his parachute after hours.
"His spore-infested parachute!" wailed Nelson-Rees.

That's why he was one of the trusted keepers of the
cells. He was a perfectionist.

Yet here was Duff, straight from a transcontinental
flight with his cancerous carryon luggage, going on about
how special these latest cells were and how careful Nelson-Rees was to be. "You're not to do anything with
these, Walter, unless you get a prior okay from us. Just
thaw them out, grow up a few separate populations, and
store them away." Jim Duff was the man who watched
over Nelson-Rees's lab for the institute. He was a natural
bureaucrat with a bureaucrat's instinct to cover his rear
flank. But he was also a friend to Nelson-Rees, who
therefore nodded politely as he sipped his vodka and
tonic.

"This is very important, Walter. There's more to it
than just science." Duff paused to let the message sink
in.

Inside the styrofoam case, within sealed plastic
flasks, floating in a red nutrient bath at room temperature, some of the cells were growing and multiplying as
the two men spoke. In a separate refrigerated compartment, duplicates of the active cells sat motionless in tiny
glass bottles, suspended in a frozen state. There were six
distinct cultures in all, a six-pack of human tumor cells,
six new weapons for "The War."

And Duff was right; they were different. For one
thing, they had come from the Soviet Union.

Richard Nixon not only dreamed of curing cancer, he
also hoped to take the chill off America's relationship
with the Russians. Toward both ends he negotiated an
agreement with Leonid Brezhnev in May of 1972 that
called for the two nations to cooperate in biomedical research. Soviet scientists were keenly interested in "The
War," particularly any recent progress that their Ameri-

can counterparts, who were thought to be at least several years ahead, were willing to share. So in November the first delegation of American cancer researchers traveled to Moscow to present their Russian colleagues with, among other things, a set of thirty viruses. There were rat viruses and hamster viruses and wooly monkey viruses and gibbon ape viruses, all of which had been found to induce cancer in these animals.

By studying viruses that caused cancer in animals, researchers hoped to find out something about analogous viruses that might cause human cancer. In fact in the early 1970s, almost everyone who counted at the National Cancer Institute seemed certain that in only a few years someone on their payroll would find the virus responsible for cancer in human beings. So far, though, all they had turned up were viruses that produced cancer in animal cells. Even those few isolated from human cells turned out to be animal viruses that had somehow found their way into human tissue but were incapable of triggering cancer there.

Undaunted, the institute pushed ahead, spending $60 million a year in the search for the human tumor virus. Competition was stiff, to say the least. For to find the virus that causes human cancer would open the way to a vaccine, a shot, like the miracle vaccination that had been immunizing people against the plague of polio for the last few decades. Cancer would be on its way to extinction, and a Nobel Prize would be the least an appreciative world could do for the scientist who made it possible. The high-pressure race had already led to several premature claims of victory in the United States. One of the most recent "winners" of the Human Cancer Virus Sweepstakes even received official congratulations from Richard Nixon after the good news was leaked to the press. Some months later the alleged agent was determined to be another animal virus—this one a mouse virus—and the race was on again.

In any case, the National Cancer Institute figured it

wouldn't hurt to give the Russians a few of these animal viruses. The collection was all done up in handsome gold lettering and presented with much pomp. It was such a hit among the Russian scientists that they felt they had to give the Americans something special in return. What was special about the cultures of human cancer cells delivered by the Russians a few weeks later was that all six had viruses growing within them, viruses that the Soviets suspected were the causes of the malignancies.

Well, as they say in diplomatic circles, it was quite something. Richard Nixon's scientific delegates were unanimously skeptical that the Russians, with their primitive equipment and lax laboratory techniques, could have come up with even one cell line that contained a genuine human cancer virus, much less six of them. No, it was too much to believe. And yet there they were, six cultures of cancer cells taken from six different patients, each culture carrying some kind of virus. Quite something. "My God, suppose they somehow stumbled onto it?" the Americans asked each other. It was simply too potentially valuable to ignore. Besides, even the purest of cynics on the American team knew there was more at stake here. These were Russians, this was detente, there were political ramifications. The Americans smiled and thanked the Russians. They packed the cells into their suitcases and brought them back home.

A couple of weeks later, Walter Nelson-Rees received a letter from the institute saying that samples of some virus-laden Russian cells, received in connection with the biomedical exchange program, would soon be delivered to his cell bank. Only four other American scientists had been given samples of the cells, and only they were authorized to experiment with them, the letter explained. As for Nelson-Rees, he was to keep some of the cells growing and to freeze the others as a safeguard against loss. The letter concluded: "No one is to be provided with these materials, or with any data acquired using these cultures, unless specific authorization has been obtained. . . . Dr. James Duff expects to visit you on January 10, 1973. He will hand carry the cultures to insure safe delivery."

Nelson-Rees thought it improbable that the viruses in these cells would prove to be human cancer agents. He was, however, pleased with himself and his lab. Once again the bigshots had come to his tidy cloakroom to check their fanciest hats.

When they had drained their drinks, Duff picked up the Styrofoam case and followed Nelson-Rees to the car. It was about nine o'clock in the evening. They drove through Oakland's harbor district to the Naval Biosciences compound. It was here, through a confusing web of agreements between government agencies and the School of Public Health of the University of California at Berkeley, that Nelson-Rees operated his cell bank. Before they entered the clean section where the cell culture work was done, where floors as well as counter tops gleamed like mirrors, Nelson-Rees asked Duff to remove his jacket and don a white lab coat. Then he handed him a pair of white nylon booties to keep the street dust on Duff's shoes from dirtying the floors.

Nelson-Rees recorded the arrival of the cultures in a big black logbook. He put the plastic flasks in a tall aluminum cabinet that looked like an oversized refrigerator but in fact was the incubator, where cells in culture grew in a tropical environment of 98.6 degrees Fahrenheit. The tiny glass bottles went into a cylindrical freezer of stainless steel—about four feet in diameter and five feet tall—that was filled with liquid nitrogen at 300 degrees below zero. Nelson-Rees slammed the lid shut, forcing a cloud of cold, white vapor into the room, and turned to Duff.

"Okay?"

"Okay."

Walter Nelson-Rees's obsessions went far beyond nylon booties. "In order to run a successful cell bank that stores thousands of different tissue cultures, that grows up and sends out as many as a dozen samples a day to researchers all over the country—to do that," Nelson-Rees often explained, "one needs to know what one is working with. One has to know which cell is which." Identification, the art and science of being able to recognize the subtle differences between cells, was his stock-in-trade. All of which is to say that he could no more leave six unidentified cell cultures in his deep freeze and incubator than Felix Unger could sneeze into his hands.

A few weeks after Jim Duff told him not to do anything with the Russian cells, Nelson-Rees disobeyed orders. He sent samples of each culture to Ward Peterson, a colleague at the Child Research Center in Detroit, who specialized in performing certain biochemical tests. This was always the first step in his routine check of anything that came into the bank. While waiting for Peterson's results, Nelson-Rees examined the cells' chromosomes, the rodlike structures inside each cell's nucleus that carry all its genetic information. After several days of gazing through a microscope at scores of slides, he had convinced himself that indeed, these cells carried human chromosomes; they were human cells. If someone had said to him at that point, "Well of course the chromosomes are human. The Russians said they were all human cells," Nelson-Rees would have nodded and replied, "Mmm, of course. But one must be sure about these things."

The cells were from human patients, all right, but he had noticed something odd. Among its many chromosomes, every human cell usually has two that determine an individual's sex. The sex chromosomes come in two varieties, one shaped roughly like the letter X, the other like a Y. If a person's cells have two X chromosomes apiece, that person is a female. If they carry one X and one Y, the person is a male. What Nelson-Rees noticed was that there were no Y chromosomes among any of the Russian cells. How strange if by chance all six cultures happened to come from women. Of course, it was well known to tissue culture experts that cells growing for a long time in a laboratory environment could simply lose chromosomes now and then. Indeed, the Y seemed particularly vulnerable to getting lost in the shuffle of division. Possibly there were some males among the original Soviet cultures whose Y chromosomes had been lost along the way; that would explain things.

Still, it bothered him.

Peterson's results began coming in about a month later. He had been analyzing the cells' enzymes, catalysts that speed up chemical reactions within the body. While some enzymes are standard equipment for human beings—nearly everyone, for instance, has the enzyme galactose-1-phosphate uridyl transferase that helps break down a sugar found in milk—certain ones appear only in small segments of the human population. These enzymes, once observed and catalogued, can be used along with other characteristics to tell cells apart.

In the first Russian cell line he tested, Peterson found an enzyme called G6PD type A, a variant of a standard enzyme that helps to metabolize glucose, another basic sugar. The form of G6PD in most people moves relatively slowly through an "electrophoretic gel," a kind of Jell-O smear hooked up to a battery or transformer. Electrophoresis indirectly measures a chemical's electric charge by testing how fast a certain voltage can drag it through this gel. The point is that in addition to the common, slow-moving form of G6PD, there is another variety that moves faster, presumably because it has a stronger charge. It was known from studies of many cells that this variant, called type A, occurs almost exclusively in black people. And even among blacks it appears in just one out of three people. Peterson's discovery of this uncommon enzyme in one of the Russian cells was not in itself unusual; there are blacks in the Soviet Union. What was unusual was his finding it in the second cell line as well, and, as the months went on, in the third, fourth, fifth, and sixth.

"It is somewhat unexpected that all six cultures would lack the male Y chromosome," Nelson-Rees said to himself. "It is even more unlikely that all six cell cultures would by chance be from blacks. And the odds that all would also just happen to have the rare form of the G6PD enzyme are . . . . Well, it's almost impossible."

He decided to notify the bureaucrats in Washington.

Jim Duff, long-distance, was upset, though not as much as might have been forecast based on his original commandment on the special handling required for these cultures. Perhaps he knew Nelson-Rees well enough to expect his friend would scrutinize them and send them to Detroit despite the warnings. Or maybe the news left him stunned. In any case, he told Nelson-Rees the observations were unusual, but they pointed to no firm conclusion. No doubt it occurred to Duff that, if true, the findings might eventually prove to be embarrassing. He was inclined not to accept them, at least not until Nelson-Rees had checked and double-checked. "There's too much at stake here, Walter," said Duff. "Don't rock the boat. Don't rock the boat until you're absolutely, 100 percent sure."

The next call went to the man at the institute who had direct responsibility for the care of the Russian cells. He was a superior of Duff's, superior not only in rank but also in his ability to worry about matters of protocol, to evaluate every development in terms of how it might offend someone, anyone, or—who knows?—maybe even endanger his job.

"What do you mean you sent the cells out for testing?"

"I sent them out for the purpose of identifying them," said Walter Nelson-Rees.

"Those cells were not to go anywhere. What has this fellow in Detroit been doing with them?"

"Nothing, just analyzing them. You needn't worry."

But the bureaucrat was worried, and one of his more intriguing worries was that the institute might get nailed

10   A Conspiracy of Cells

for violating federal quarantine procedures. It was awkward enough that Nixon's delegation had smuggled the Russian cultures into the United States in the first place. Being bathed partly in bovine serum, the cells were a potential source of hoof-and-mouth disease, a scourge then rampant in the Soviet Union. Authorities with the U.S. Department of Agriculture had become quite agitated when they heard about the cells' unofficial entrance into the country. They had calmed down slightly when the cancer officials permitted inspections of the few labs authorized to work on the cultures. But now, if news leaked out that Nelson-Rees had been shipping the stuff wherever he pleased . . . .

"Doctor Nelson-Rees, you were told specifically not to distribute those cells. How could you have sent them to Detroit?"

"I've just told you. I sent them to Detroit to be identified."

By the time the conversation had circled back around itself a few times, it was clear to Nelson-Rees that the bureaucrat was not interested in hearing about his findings. He then called Wade Parks, an institute virologist and one of the privileged researchers allowed to examine samples of the Russian cells. Parks said he had tentatively identified the virus from one of the cell lines. Unfortunately, it appeared to be a monkey virus—another confounded animal virus. This one, Parks said, had never been known to do anything related to cancer or to be anywhere interesting. Not only that, preliminary checks of the other cell cultures suggested they might all be carrying the same worthless virus. The implication was that the Russian cells were duds. The elusive human tumor virus, it appeared, remained elusive.

But Nelson-Rees was more interested in the cells themselves than the viruses they held. What with his own findings and now this from Parks, there was something obviously strange about these cell lines. All lacked the Y chromosome, all carried the rare enzyme, and now all contained the same virus. The theory taking shape in his mind was both obvious and preposterous; he had to try it out on someone. "Has it ever occurred to anyone that these might be all the same cells?" he asked the virologist.

He would not have been surprised to hear Parks dismiss the idea. After all, how could they be the same cells? Each culture had come from a different research institute in the Soviet Union. Each was seeded by a bit of tissue taken from different patients with different kinds of cancer. How could they now be identical? And what kind of cell would they all be? Parks, however, said none of this. Instead he said that the same thought had occurred to him.

Nelson-Rees couldn't say—not precisely, anyway—how all the cell lines had become one. Nor could he say what kinds of cells were in the six cultures originally. But he was almost sure of what was growing in them now. He even knew the name of the woman to whom the cells, all of them, belonged. But he needed proof.

As it happened he and an assistant had just learned a technique that could help him get the proof. It was a method of applying purple stain to a cell's chromosomes. The chromosomes absorbed the color in selected areas and ended up looking like barber poles. Instead of neat stripes of red and white, though, they had bands of purple running across them at irregular intervals. The banding patterns were unique to each chromosome, like fingerprints, which made cells eminently more identifiable. If a researcher knew the banded fingerprints for various cell lines, he could use them to identify an unknown cell with far more certainty than by merely determining the lack of a Y chromosome, for example, or the presence of a rare enzyme.

Nelson-Rees and his assistant spent several months preparing the Russian cells, staining the chromosomes, photographing the fingerprints, and checking them against the fingerprints of their prime suspect. When they were through, the conclusion was inescapable. Maybe the bureaucrats didn't want to hear it, but they got another phone call. It was bad news, reinforcing Parks's impression that the cells were probably worthless to the quest for the human tumor virus. But it went beyond that.

These were not distinct cultures of cancer cells from six different patients in the Soviet Union. They were all the cells of an American who in her entire life had probably not been more than a few miles from her home in Baltimore, Maryland.

A housewife with four children, this woman had been stricken with cancer at the age of thirty. She died in 1951, more than twenty years before a group of smiling Russians proudly presented a group of smiling Americans with six of their most promising cell cultures.

Her name was Henrietta Lacks.

2

The Seed That Took

Who? According to the medical records, Henrietta Lacks was born somewhere in Virginia on 18 August 1920, the daughter of John and Eliza Pleasant. Her mother died while delivering her tenth child; Henrietta was very young at the time. Under the section titled "Personal and Social History," the records say that Henrietta was the mother of four and that she had a happy home life. Her major anxiety was due to a ten-year-old epileptic daughter who could not speak. As for education, Henrietta had gone as far as seventh grade. She drank beer only occasionally. Her husband David worked at the Sparrow's Point shipyards and earned an adequate income. They rented a five-room house with coal heating on New Pittsburgh Avenue in Baltimore.

Under "Past Health" the description is medically unremarkable: As a child she had chicken pox, measles, and periodic trouble with her tonsils. As an adult she had occasional headaches and often suffered a stuffed-up nose, probably the result of a deviated septum. In short, the records describe a young woman who had been quite well and relatively free of cares until January of 1951, when she noticed a pink discharge spotting her underclothes.

By 1 February, the day Henrietta Lacks made her way through the winter rain to the women's clinic of the Johns Hopkins Hospital, the discharge was blood red.

The gynecologist who examined her was baffled. He quickly found the source of the blood; it was a puffy lobe of tissue about an inch in diameter on the left side of her cervix. But as to what it was exactly, this experienced physician couldn't say. All the cervical tumors he had encountered were pale, almost white, because they lacked an ample blood supply. In fact most were so starved for blood that they developed ulcers. Not this thing; it was shot through with vessels and so full of blood that it seemed to glow a deep red. There were no signs of ulceration either. It was a mean and hardy thing, whatever it was.

The gynecologist took Henrietta Lacks across the hallway to the venereal disease clinic, thinking it might be a lesion caused by syphillis. But the technicians there could find no trace of syphillis bacteria within her cervix. Still puzzled, the gynecologist cut off a small section of the lump and asked a pathologist to examine it under a microscope. Henrietta Lacks received the verdict that same day: an unusual form perhaps, but this dark red growth of tissue was unquestionably cervical carcinoma, a malignant tumor of the cervix.

Eight days later she returned to the hospital for her first session of radiation therapy. A surgeon stitched a tube containing radium capsules to the wall of her cervix. It remained there for twenty-four hours, bombarding the tumor and the rest of her abdomen with high-energy gamma rays. After a second operation to remove the tube, the surgeon wrote his report: "Patient feels quite well tonight. Morale is good and she is ready to go home." His records also note that during the earlier operation, he had sliced off two tiny pieces of cervical tissue—one normal, the other from the tumor—and had given them to a researcher in the Hopkins medical school named George Gey.

George and Margaret Gey were husband-and-wife pioneers in the field of tissue culture. George, who stood well over six feet tall and had a broad chest, was an idea man and a tinkerer. Among his many inventions was something called a roller tube, a sealed glass vessel in which living cells were periodically bathed in a nutrient solution as the tube was slowly rotated. He blew the glass for the first tube himself and rigged up a primitive rotor using the pendulum of a clock. It became a standard in the field. A zealot when it came to the care and feeding of cells, George Gey once bought an entire boxcar full of powdered soap for washing culture dishes; it seems the local market had switched to a newfangled detergent that sometimes left residues harmful to cells, and George wanted to make sure he'd never be forced to use the stuff. Although a native of Pittsburgh, he had a country boy's air. He spoke plainly and with warmth as readily to cab drivers as to scientific dignitaries. He ran his lab like an informal college of tissue culture, sharing equipment and knowledge with any student or scientist who wanted to learn. And for a few hours every Wednesday, he went fishing.

Margaret, trained as a surgical nurse, was the meticulous director of day-to-day operations in the laboratory. As long as everyone did his share, she was an amicable boss. If necessary, however, she could play the role of the stern head nurse. She saw to it that the cultures were fed on schedule, that the glassware was sterilized according to specifications, that the records were kept in order. George was regularly out of the lab, hunting for funds and lecturing, but Margaret was always there, working long days and weekends, all of it without pay.

Together the Geys were trying to put some science into the folk art of growing cells. If reliable methods could be devised to keep cells alive in a roller tube and to trick them into thinking they were still inside a body, then scientists could learn firsthand about human biology without having to experiment on human beings directly. The workings not only of healthy cells but also of sick ones could be revealed. That possibility—of capturing human diseases under glass and dissecting them for their underlying causes—was what lured the Geys and several other research teams into the business of culturing tissue. Their greatest hope was to establish and study long-lasting cultures of the most dread human disease, to have, as some put it, "a tumor in a test tube."

But it was slow, difficult, sometimes grizzly work. In order to give their cells a proper feeding solution, for example, they had to find the raw materials themselves. Several times a week Margaret went to the Hopkins hospital's maternity ward to collect one ingredient believed to stimulate cell growth: blood from human placentas. One of the maternity nurses would press a button that set off a buzzer in the Geys' lab, signaling that a fresh placenta was being put aside for them. Margaret soon arrived, cleaned off the sac's umbilical cord, and, plunging a fat syringe into one of the cord's larger blood vessels, pulled out as much as 50 cubic centimeters, about a third of a cup. Back at the lab the serum component, the slightly yellowish fluid containing proteins, was removed from the blood and combined with what the Geys called beef embryo extract. This was the ground-up remains of a three-week-old cattle embryo, collected periodically from a cooperative packing house. Because the recipe also called for chicken plasma, every so often the Geys visited a nearby poultry factory. They usually went about dawn so that few workers would be around to witness the spectacle. Margaret's job was to pull back the wing, swab the area around the ribs with alcohol, and hold the animal still while George poked the syringe directly into its heart. They took 50 cubic centimeters from each chicken, and most walked right off the table and back to the yard. They had a deal with the owner to buy any bird that didn't survive. On such occasions Margaret cooked the unlucky animal for dinner that evening.

As for the cells to be grown in this elixir of plasma

and serum and embryo mash, the Geys had to go out and collect them as well. Every day George scanned the hospital's list of upcoming operations and procedures, looking for interesting sources of tissue. He or Margaret or one of the technicians stood in the appropriate operating room holding a few petri dishes ready for a scrap. Then they raced back to the lab, placed the pulpy fragment on a clot of chicken plasma, added the remaining parts of their cell food, and hoped that it would take.

It was discouraging. No matter how clean their glassware, no matter how potent their nutrient solution, no matter how careful their technique, cells simply weren't comfortable growing outside the human body. The Geys had had considerably more success with animal cells, including some that had survived for years now. But most human cells shriveled and died right away. Some held on for a few weeks, coaxed and coddled by Margaret's diligence, and then gave up. It was a testimony to the Geys' abilities that by early 1951, they had managed to keep a few lines of human cancer cells alive for several months. It was a testimony to their determination that they kept trying to establish new ones.

Mary Kubicek, however, was running low on determination. A twenty-one-year-old technician just out of college and newly trained by the Geys, Mary was frustrated by her current assignment: an attempt to establish a culture of cervical cancer cells. Mary was shy and slightly insecure; that made her doubly careful and especially hardworking, even measured by the high standards of the Gey lab. Nonetheless after trying samples of cervical cancer from a dozen different patients, all she and her fellow technicians had to show were dead cultures. When George Gey announced around noon on 9 February that he had dropped off yet another cervical cancer biopsy in Mary's work area, she didn't attend to it right away. Uncharacteristically, she lingered a few minutes at the lunch table and finished her sandwich. What's the difference, she thought. It'll just be another useless attempt.

The tumor fragment was red, roughly square, less than half an inch on a side. Gey explained that it had been taken from a patient in the women's clinic just before she underwent radiation therapy. Following the usual procedure, the tissue was code-named with the first two letters of the donor's first and last names: HeLa. Mary cut away the decaying, brown tissue around the edges and sliced the remaining chunk into tiny cubes. She pipetted a few drops of chicken plasma into several roller tubes and placed four cubes of tissue in each tube. She waited five minutes for the plasma to stiffen into a clot, and then added the rest of the feeding solution. Finally, she placed the tubes into a rolling rack inside the laboratory's incubator.

Had Mary been staring through the glass doors of the incubator, she wouldn't have seen the early signs of life. It happens so slowly at first and so sporadically. Maybe it was a matter of hours, maybe the better part of a day passed. At some point, the cells in each little island of tumor began to quiver and dance . . . and multiply. Where there had been one, there were now two. Where there had been two, now four, now sixteen, now thirty-two. In a few days, the signs of growth were visible: around each cube a translucent band of new cells was taking shape. On the fourth day Mary had to remove the burgeoning bits of flesh from the roller tubes, carve them up into smaller pieces, and transfer the cuttings into additional tubes.

Every other human cell line in the Gey lab had eventually weakened and faltered. Yet as the months passed the HeLa cells showed no such vulnerability. They just kept growing, doubling their number every twenty-four hours "spreading like crabgrass!" was how Margaret Gey described it. At one point George Gey compared the performance of the cancerous HeLa cells to the performance of the normal cervical tissue taken from Henrietta Lacks: the tumor cells were growing ten to twenty times faster. It was too early to start celebrating, and George Gey wasn't the sort to stop and congratulate himself anyway. But there was no doubt about it, these cells were different.

The tumor within Henrietta Lacks was different too, though the doctors didn't know it at the time. Most cervical tumors—especially those found at an early stage, as was Henrietta's—were easily beaten back by radiation. Most patients were still alive five years after therapy. So after several more radium treatments, the doctors gave her one month of X-ray therapy and hoped that would be the end of it. "No symptoms referable to the pelvis," wrote one who examined her in May. "Cervix is normal in size, mucosa red and smooth, cervix freely movable. Good radiation result. Rx: Return in one month." In June he wrote: "Patient feels fairly well, but continues to complain of vague lower abdominal discomfort. Cervix appears perfectly normal. No evidence of recurrence. Rx: Return in one month."

By late July Henrietta's side aches could no longer be described as vague. Now they were extreme and radiating down into the groin. Not only that, something was constricting the area around her bladder so severely that her kidneys began to swell with urine. The doctors also found a large, stony mass of tissue on the inside of her pelvis and another tumor in a lymph gland. With shocking speed the cancer had reappeared and spread so extensively that the doctors considered it inoperable: "For this reason, we are giving the patient a second course of deep X-ray therapy purely as a symptomatic therapy."

She was admitted to the hospital on 8 August, ten days before her thirty-first birthday. For the next twenty-two days she ran a constant fever of 100 to 102 degrees; she also began vomiting regularly. Despite the X-ray treatment, tumors were filling her abdomen. The treatment was halted.

In mid-September, with the fever, pain, and nausea continuing, she developed uremia, a buildup in the blood of poisonous waste products normally eliminated by the kidneys and bladder. The doctors tried to insert a catheter tube through the mass of tumors that choked her bladder, but failed. They gave her transfusions, replacing her own polluted blood with a fresh supply. Her intestine, however, was also blocked and her abdomen was beginning to expand.

On 26 September one of the doctors looked over her order sheet, the long list of various drugs, procedures, and treatments that had been prescribed to try to save Henrietta Lacks. At the bottom of the list, he scrawled: "Discontinue all medication and treatments except analgesics." All he could do was try to relieve the pain. On 29 September, Henrietta Lacks became disoriented, apparently confused about where she was and what she had been going through. She stopped breathing at fifteen minutes after midnight on 4 October 1951.

It had been only eight months from the time the small red patch was first discovered in her cervix to the day it killed her. For cancer of the cervix, the doctors said, it was some kind of a record.

No one in the Gey laboratory laid eyes on their unfortunate benefactor until 4 October when George Gey noticed a listing for the autopsy and went to observe. Because he wanted a few more samples of the remarkable tumor cell, he asked Mary Kubicek to meet him there and collect them.

The Hopkins autopsy room was buried in the basement of another building. Mary had never been there before. In fact she had never been to any autopsy facility or to a morgue or to anything like that. She made her way anxiously through the maze of dim underground hallways that connected the laboratory building to the other basement.

The room had high ceilings and a bare stone floor. At the far end Mary saw a body on the table. A pathologist was hunched over it, at work, and Gey stood nearby. The dead woman's arms had been pulled up and back so that the pathologist could get at her chest. Even from a distance Mary could see that the body had been split down the middle and opened wide.

She walked to the table, sidestepped one of the outstretched arms, and held out her petri dishes. As she waited she gaped at the greyish white tumor globules that filled the corpse. It looked as if the inside of the body was studded with pearls. Strings of them ran over the surfaces of the liver, diaphragm, intestine, appendix, rectum, and heart. Thick clusters were heaped on top of the ovaries and fallopian tubes. The bladder area was the worst, covered by a solid mass of cancerous tissue. "Bladder pushed to anterior abdominal wall," wrote the pathologist, "Almost entirely replaced with tumor."

Mary's eyes wandered down toward the corpse's feet and suddenly she was overcome. The toes. They were painted with bright red nail polish, and a dainty job it was. It suddenly made this carved-up cadaver real. All the laboratory experimentation had never hinted at the tragedy of this disease. But here, she thought, over here on the table is the proper demonstration. Here is what cancer does.

The pathologist sliced pieces of tumor from different organs and dropped them one-by-one into the dishes in Mary's hands. It seemed to Mary that he took forever. Finally, she fled: across the expansive stone floor and out of the autopsy room, through the catacombs, and up the stairs from the basement. Back at the lab she concentrated on the task at hand, cutting up the tissue and placing the pieces into roller tubes. In the bright familiar surroundings, her horror quickly faded. The image of the delicately polished toenails, however, lingered. In the years that followed, that sight came back to her often.

As for the cadaverous cells, they would not grow. The uremia had made it impossible for anything to live within the body of Henrietta Lacks. The cancer's sabotage had been so effective, it killed not only its host, but also itself.

That's not quite right, of course. Part of what had been Henrietta Lacks's cancer was not dead. Some of the cells had escaped their own poisonous wreckage eight months earlier on the edge of a surgeon's knife, abetted by the tissue culturing skills of Mary Kubicek. Dining on clotted chicken plasma, chopped beef embryo, and the blood from human placentas, the surviving cancer cells of Henrietta Lacks were living quite comfortably—thriving, in fact—in glass tubes in George Gey's lab.

3

HeLagram

In the early 1950s many Americans would have named cellophane science's niftiest invention. To the small community of researchers trying to grow human cells in their laboratories, however, the truly great breakthrough was the HeLa cell. At last, here was a cell with staying power, the first durable piece of a human being that could be watched up close and tinkered with. No more racing to finish an experiment before a sickly culture wheezed and fell dead. This cell would last, not just through one series of tests but through dozens, for months, maybe for years.

In fact George Gey and two co-workers at the University of Minnesota showed that HeLa cells were so rugged they could survive a 2,500-mile trip through the mail. They sent twenty-nine live cultures by air, rail, and truck from Minneapolis to Norwich, New York, and back; all but one returned in fine health. A cell line that held on in the lab was what everyone had been hoping for, of course, but one that could endure handling by the U.S. Post Office was a blessed miracle.

Soon it seemed every biomedical scientist in the country was either sending or receiving a HeLagram. Gey started it by mailing samples of HeLa cells to a few close colleagues, who grew up some extra cultures and sent them to their friends, who did likewise. When the frenzied demand for HeLa outpaced this informal network, a number of laboratories set up full-scale production lines and began passing around HeLa cultures the way McDonald's shovels out its burgers and fries. Mary Kubicek heard that one shipment of HeLa was being carried by backpack into Chile and another was on its way to Turkey. In a few years HeLa cells would even travel into space aboard the *Discoverer XVII* satellite.

Cancer researchers craved HeLa most of all because it was their long-sought tumor in a test tube. They watched the cells react to a battery of toxic chemicals and photographed the weirdly shaped chromosomes. Scientists interested in the general workings of human cells clocked the rate at which HeLa cells multiplied and studied their production of proteins. Virologists, who found that polio viruses multiplied a millionfold just two to three days after infecting a HeLa culture, made HeLa their major new tool for studying the viruses. A year later they knew enough about polio to produce the first successful vaccine.

And, of course, every scientist who wanted to learn the methods of tissue culture insisted on starting with HeLa. The newcomers had been told how frustrating much of the work would be, how rarely a piece of tissue would yield a sustainable cell line. But with HeLa, they knew they couldn't lose. The advent of HeLa was not only a boost for the beginners, though. It also seemed to change the luck of the researchers who were struggling to establish other human cell lines.

It was as though George Gey had broken biology's four-minute mile. Once he had shown it was possible to produce a strong and long-lived cell line from human tissue, his fellow cell culturists went back to the lab with new confidence and enthusiasm—and damned if they weren't able to do it too. One scientist successfully cultivated a hardy strain of human liver cells. Another started up a cell line from a bit of amniotic sac. Then came a culture from a tumor of the larynx, followed quickly by sturdy lines of embryonic kidney cells and adult heart cells and cancerous blood cells. True, the cultures didn't always bloom right away. But eventually, a few days or a few weeks after being seeded, somehow they all began growing at a strong and steady pace. For the next ten years, from the early 1950s to the early 1960s, the science of cell culture flourished as well.

A few of the old guard saw the calamity coming. With so many new cells in circulation, and with no means of positively identifying them, they knew there would be mixups. So they tried to head off the problem. Some learned to tell cells apart by the shapes of the chromosomes, others by how antibodies reacted to particular cultures. The methods were crude, but they worked well enough to show the scientists they had been right to worry.

One of the first cultures they checked was a human line that one day, for no apparent reason, lost its susceptibility to polio. Their identification tests suggested the cells were no longer human but had somehow been replaced by mouse cells. Then the same thing happened to a monkey cell line. Soon they found human cells growing in what should have been pig, duck, and mouse cultures. And there were mouse cells growing in rabbit cultures and rabbit cells where monkey cells should have been.

Most cell culturists were still celebrating the renaissance. But by the late 1950s many samples of the useful cell lines—both the old, established animal cultures and some of the new human ones—had lost their identities. An investigator who needed a normal monkey cell for his experiments could no longer be sure he wasn't working on a tumor cell from a human larynx. How could he interpret results when he couldn't even say what he had been experimenting on?

The reasons for the mess were obvious to the old guard. In a word these new people were sloppy. They were probably mislabeling cultures. Hell, they were probably using the same pipettes to feed different cultures, inadvertently picking up a few cells out of one dish and dropping them into another. If the first cells were stronger, they would grow over the second culture and replace it. Such faux pas slipped by unnoticed in many hectic labs where accurate record keeping had become a lost art. In addition, cell swapping was rampant. And when one researcher traded materials with another, he also traded mistakes.

Well, things simply couldn't go on this way, declared the veterans who had uncovered the confusion. They decided to rescue the field by setting up a central cell bank, a Fort Knox for cell cultures. Not just any cell cultures, you understand, only those with clearly defined characteristics and carefully documented histories. No shadowy pasts allowed, no vagabond cells that had wandered from one nameless lab to another. With the help of the National Cancer Institute, which was also beginning to worry about the quality of cell cultures, the group announced in 1962 that the nation's "reference cells" would be housed at the American Type Culture Collection, a private supplier of biological materials in Washington, D.C. Workers there would maintain these purebred cultures and distribute them to any researcher who wanted the very best. Over the next four years, the cell bank filled its refrigerators and incubators with more than two dozen high-quality cultures. It looked as if the science of tissue culture had backed away from the brink of chaos.

Yet as careful as they were, the founders of the bank were for years haunted by doubts. Most of their screening tests determined only what kind of animal a cell line came from. They could tell mouse from human, but they couldn't easily tell most human cells apart. What they needed were markers, chemical or structural fingerprints that were readily recognized and unique to each cell line *within* a species.

In 1966 a Seattle geneticist named Stanley Gartler stood up at a scientific meeting in Bedford, Pennsylvania, and offered the tissue culturists just what they needed. In appreciation, the tissue culturists practically ran him out of town.

As part of his study of human genes, Gartler had been looking for long-lasting cell lines that produced certain isoenzymes, enzymes that are present in every human cell but may vary in style from person to person. One of the isoenzymes of interest to Gartler was G6PD, the glucose metabolizing enzyme that comes in two styles, type A and B. Another was PGM, available in types 1, 2, and 1–2, a combination of styles analogous to the blood type AB.

Gartler first analyzed the PGM in a few of the established human cell lines and was surprised to find the same form of the enzyme in each: type 1. He tested a few more and then a few more. He stopped at eighteen. Now it is true that in an average population many people would have type 1 in their cells, but the statistics require that nearly a third of them should not. That was what bothered Gartler. When he tested the cultures for the G6PD enzyme, again they were all the same: type A. That meant that every cell line he checked had come from a black person. Most of the established cell lines, however, had reportedly been cultured from Caucasian patients. Only HeLa was known to have come from a black.

Because HeLa was the earliest successful cell line, used in virtually every lab before the rest of these eighteen cultures came along, Gartler drew what he thought was an obvious conclusion.

Oddly enough he didn't think his serendipitous finding was all that important. After all, it offered no new scientific principle or insight. The main point to him was that these variants of different enzymes could be used for telling human cells apart; they were practical tools for sorting out existing mix-ups and preventing future ones. In fact the title of the paper he submitted to the Bedford conference was "Genetic Markers as Tracers in Cell Culture." But then Stan Gartler was a geneticist, not a tissue culturist.

"My God, they're going to tear you limb from limb," said a friend when Gartler arrived at the meeting to deliver his report. "I can't believe what you're saying."

Gartler said it nonetheless. He stood up in the convention hall of the Bedford Springs Hotel and said that his tools for distinguishing human cell cultures demonstrated that most of them weren't different at all. He said that the eighteen cell lines he tested, samples of which were now in the vaults of the nation's new cell bank, were really just the ever-popular HeLa cells. He added, almost incidentally, that researchers who had experimented on these cervical cancer cells believing they were liver, or blood, or bone marrow, or anything else had better reconsider their findings. "The work is open to serious question," said Gartler, "and in my opinion would be best discarded."

The tissue culturists were not pleased to hear this, particularly from a geneticist, particularly since they had spent the last ten or fifteen years studying samples of these cells, thinking they were looking at many distinct forms of cancer and at normal tissue from lots of different organs. Even the founders of the cell bank who had suspected trouble found it hard to believe. And to the researchers who had actually created the cultures on Gartler's list of spoiled goods, who had toiled for years and suffered repeated disappointments before they finally got those cultures to take root, his findings were impossible. They began hurling skeptical questions.

Just how did he know the cells hadn't been taken over by HeLa in his own laboratory?

They were all analyzed as soon as he received them, Gartler answered.

But hadn't some been sent to him frozen?

We can analyze frozen samples, said Gartler.

Well, isn't it possible for a cell to change its enzyme type—from G6PD type B to A, for instance—and still remain the same in all other ways?

Gartler sighed. These people were obviously not geneticists. He explained that there are indeed some characteristics in a cell that may readily change. In a developing embryo, for example, a cell that had been relatively nondescript may, in the process of "differentiating," turn on various genes that manufacture different kinds of proteins. But there is only one gene in each cell that controls the production of the G6PD enzyme; that gene specifies either type A or type B, and it is always turned on.

What about random mutations?

Gartler estimated that the chances of a mutation reshaping the gene for type B into a gene for type A would be less than one in a billion. Even if so unlikely a mutation took place in a single cell, why would that single cell overtake the rest of a culture, Gartler asked. And why should such unlikely occurrences have happened in all eighteen cell lines?

Leonard Hayflick stood up. A highly respected cell biologist and an officer of the Tissue Culture Association, which was sponsoring the conference, Hayflick was also the originator of a cell line known as WISH. WISH was one of the cultures that Gartler had discredited. It was a culture grown from a scrap of amniotic sac. In fact, Hayflick told the attentive audience, WISH had been taken from the amniotic sac in which his daughter Susan came into the world. WISH stood for Wistar Institute, where Hayflick was working, and for Susan Hayflick. Since Hayflick and his wife were both Caucasian, the claim that WISH showed a genetic trait found only in blacks was a tad awkward.

In perfect deadpan, Hayflick announced, "I have just telephoned my wife, who assured me that my worst fears are unfounded."

The crowd thought that was hilarious, and the tension eased. But Gartler wasn't so sure that Hayflick was genuinely jovial. And when the laughter died down, the eminent cell biologist dismissed the geneticist's conclusions, saying simply they were very difficult for him to accept.

Then rose Harvard University's Robert Chang, another luminary of cell biology and a trustee of the Tissue Culture Association. Chang was the creator of one of the most popular of human cell lines. The Chang liver culture was used extensively in studies of liver function.

Whatever culture Gartler claims to have analyzed, said Chang, it wasn't a culture that came from Chang. "I have never sent him any cell line, and I don't remember ever having corresponded with him."

Gartler explained that one of his samples of Chang liver had come from a co-worker at the University of Washington in Seattle. The other came directly from the cell bank at the American Type Culture Collection. In fact, six of the eighteen cultures he examined had come straight from that storehouse of only the best and most carefully screened cultures. The important point, said Gartler, is that while there may well be some genuine samples of these cultures at certain laboratories, there are others in active use that are impostors. Unless experimenters can tell the bona fide from the bogus, he said, much of the research done on these cultures is in doubt.

It looked to Gartler as if Chang was contemplating murder—or suicide—but all he did was sit down.

More skeptical questions, more icy speeches. The session finally ended at noon, Gartler escaped from the room, and the tissue culturists changed their tactics. Now it was a war of isolation. They ostracized him for his wild and insolent claims. Through most of lunch he sat alone. One of the few who dared to join him was a young researcher from California. As a technical advisor to the cell bank, the researcher was as disturbed as the others about Gartler's findings. At the same time, though, he admired Gartler for sticking his neck out, and he told him so. The researcher's name was Walter Nelson-Rees.

4

Out of Thin Air

It wasn't until two years later that Gartler's incredible conclusions were confirmed. By 1968 two independent research teams had applied his methods to all the human cultures deposited in the cell bank at the American Type Culture Collection. Out of thirty-four cell lines, they found twenty-four to be HeLa.

How could it have happened?

The most likely explanation was the same combination of sloppiness and opportunity that in the late 1950s had shuffled mouse cells with human cells and duck cells with monkey cells. HeLa had a number of advantages that helped it pull the trick off on a disastrously large scale. Since it was the first useful human cell line, it was the most ubiquitous: wherever technicians were careless, there were always a few HeLa cells nearby to take advantage. Because it was also one of the most vigorous cultures known, it could easily take over weaker cultures if given half a chance.

HeLa was so tenacious, in fact, that it probably didn't need to wait for a lab worker to use the same pipette on different cultures. A startling series of experiments reported in 1961 by Lewis Coriell, one of the old guard who had helped start the reference cell bank, had shown HeLa could literally appear out of thin air. Coriell, working at the Institute for Medical Research in Camden, New Jersey, found that merely pulling a stopper from a test tube or dispensing liquid from a dropper could launch tiny airborne droplets containing a few HeLa cells. When the drops landed on open petri dishes holding live cultures, the HeLa cells began growing so feverishly that in three weeks they overwhelmed the original cultures.

To some it had seemed an unbelievable observation. But now it looked as though much the same thing must have happened to all the human cultures that appeared in the 1950s soon after HeLa. Those cells had probably been as weak and hard to cultivate as the ones that had been tried in the pioneering days; they were easy victims for HeLa. A few of the more cynical scientists suspected there had never been anything but HeLa in those new cultures. The cells from the original tissue samples had probably died immediately, leaving a culture dish of nutrients ripe and ready for the next HeLa cell that happened by.

Either way, for Coriell and some of the other experts, HeLa's surreptitious spread explained a few puzzling observations that cancer researchers had made in recent years. One such enigma was "spontaneous transformation," a mysterious process by which benign cells suddenly turned malignant. There it was right in the dish, the very nut of the cancer problem: healthy cells, going along in a calm and orderly way, abruptly burst into unbridled growth. Not only did they grow faster, they were no longer bound by the normal cells' lifetime limit of fifty to sixty divisions. The transformed cells ignored their biological clocks and continued doubling without end.

The weird thing about spontaneous transformation was that until the late 1950s and early 1960s, it was never known to occur in human cells. Cultures of rat, mouse, hamster, and other animal cells had been spontaneously transforming for years, but never a human cell culture. Then suddenly it was happening all the time. Not only that, these spontaneously transformed cells grew rings around many cells taken directly from patients' tumors. Nobody could explain why normal cells that turned malignant in the lab should be more aggressive than cells that had become malignant while in the body, but for the moment scientists were delighted to have all these tenacious new cultures.

Much later it became clear that these transformations were not spontaneous at all, but had been triggered by outside agents. In the case of the nonhuman cells, chemicals in the nutrient medium, oxygen in the air over the cultures, even fluorescent lighting in the laboratory were eventually found to inflict genetic harm that can turn normal cells cancerous. As for the human cells, most "transformations" appeared to be nothing more than takeovers of the cultures by the feisty HeLa cells.

Spontaneous transformation was not the only myth HeLa created about the nature of cancer. Researchers had observed that cancer cells shared many fundamental characteristics, and there had begun to emerge a unifying theory: all cancer cells grew relatively quickly and had the same basic nutritional requirements; they seeded new tumors when inoculated into the cheeks of hamsters; many had abnormally shaped chromosomes; and most carried the same surface antigens, proteins on the outside of the cell that stimulate the body's immune system. Like winning lemons in a casino full of rigged slot machines, these traits kept coming up one after another in dozens of cell lines the scientists thought had come from dozens of cancer patients. The truth was they had been studying one line of cells masquerading as all the others, and the common traits they saw were those of a single tumor, the one that killed Henrietta Lacks.

"They described a lot of things they thought were being produced by intestine and kidney and other cells," recalled Coriell years later. "There was a lot of data in the literature that was just wrong, just a lot of wasted time because they were all working with HeLa."

Cyril Stulberg of the Child Research Center in Detroit, like Coriell one of the deans of cell culture, came as close as anyone to assessing HeLa's effect on this early period of cell biology and cancer work. He made the remark in a letter to Nelson-Rees many years after Gartler's HeLa revelations. "I didn't realize then what the succeeding years would bring," Stulberg wrote. "Naturally, at the time, I was very defensive because I saw 15 years work go down the drain."

But Stulberg, Coriell, and the other veterans faced up to the calamity and began to clear away the wreckage. As the architects of the fledgling central cell bank, they were reluctant to simply throw out the twenty-four HeLa-contaminated cultures. HeLa, after all was a sturdy line and these individual strains had various quirks and characteristics that made them particularly useful to certain fields of research. The solution, they decided, was an addendum to the cell bank's catalogue in 1968 warning that the twenty-four lines were actually HeLa and should be used as such. Moreover, they required even more detailed descriptions of new lines deposited in the bank and recommended not only Gartler's enzyme tests but any other promising techniques of identification as well.

This time, they hoped, they would put HeLa contamination and all the other chaos behind them. And, indeed, as the 1960s came to a close, it appeared that the cell biologists had recovered from Stan Gartler's bombshell. Gartler himself went back to studying genetics, having done quite enough for cell culture.

As for what made HeLa so damned stubborn in the first place, the scientists could only speculate. In 1970 two researchers dug out the original biopsy slides of Henrietta Lacks's tumor and uncovered one possible answer.

The pathologist at Johns Hopkins in 1951 had described her illness as the most common type of cervical cancer, involving "epidermoid" cells, which form a skinlike covering over the cervix. He had made a mistake. It was clear to the researchers who reviewed the slides twenty years later that the tumor was made up of "immature glandular" cells. Such tumors were thought to be far more aggressive in the body. That would explain the speed with which Henrietta Lacks's cancer killed her. Maybe HeLa's widespread contamination of other cell lines demonstrated that immature tumors are aggressive in culture as well.

Another theory held that though HeLa started out as a pretty scrappy scavenger of petri dishes, years of breeding in laboratories all over the world had made it even scrappier. As investigators subjected HeLa cultures to different stimuli—X rays, for example, and viruses— single cells must have changed, some becoming still stronger, others turning delicate. The fragile variants would have died off, leaving only those most fit for life in the lab. A corollary of this theory was that new strains of HeLa would have emerged with personalities the original HeLa cell never had. Though it wasn't yet generally known, there were indeed new subfamilies of HeLa cells making the rounds.

One of these novel strains was quietly touring Europe in the late 1960s and early 1970s. It had started as a standard HeLa culture, probably sent by an American scientist to friends abroad, who then passed it around. At one of the labs it visited, this HeLa culture contaminated a culture of monkey cells, killing off the monkey cells but inheriting a colony of viruses that had infected them. It first surfaced in 1973 when a group of West German scientists reported the surprising presence of monkey viruses in a human cell line. But by then, it had traveled extensively, no doubt hopping in and out of culture dishes under a variety of aliases. When researchers tried to reconstruct the wanderings of this HeLa variant many years later, all they could say with certainty was that it must have slipped into the Soviet Union at some earlier point—because it was this strain of HeLa cells growing in six separate cultures that the Russians handed back to the Americans in 1972.

5

In the Purple Palace

It wouldn't be right to say that the officials of the National Cancer Institute were surprised when Nelson-Rees delivered the bad news in the spring of 1973. Given the sorry state of Russian research, it came as no shock; that was the scientific point of view. Being bureaucrats, however, the institute people saw this as a political mess fraught with ironies that would delight the press and other antagonists. It was bad enough that the earliest fruit of Nixon's scientific détente was the discovery that the Russian cultures were good for nothing. But for the cultures to have been spoiled by the cells of some posthumously hyperactive woman from Baltimore, cells that had probably slipped into the Soviet Union as the result, however indirect, of some American researcher's largesse many years earlier—that was too much. No, they were not surprised. They were mortified. If true, the news of Henrietta's Russian appearance would be a diplomatic nightmare.

Well, then, maybe it wasn't true.

They urged Nelson-Rees to check his results. They cautioned him not to say a word to anyone. They suggested this could jeopardize the future of the entire biomedical exchange with the Russians. And reminding him that he had broken the rules by sending the cells to Detroit without permission, they extracted a written apology from him. But along with the apology Nelson-Rees sent the evidence, summaries of the many tests and retests. By early fall, the bureaucrats had no choice but to begin quietly discussing how best to break the news to the Russians. The Soviet delegation was due to visit the institute's headquarters in Bethesda, Maryland, in a few months, just as the American entourage had visited Moscow a year earlier. One of the main events scheduled was a review of what the Americans had learned about the Russian cell cultures.

In October Nelson-Rees received a phone call from John Moloney, assistant director for viral oncology at the institute and the leader of the American delegation to Russia. Never before had Nelson-Rees received a phone call from Moloney. Moloney was just one step down from Frank Rauscher, the director of the institute. And Rauscher, unlike the directors of other U.S. health institutes, was appointed by the president himself. It was a little like a sixth-century English footsoldier getting a call from Sir Lancelot.

Moloney said he would appreciate it if Nelson-Rees would come to Bethesda to make a presentation to both institute officials and the visiting Russians. The topic would be the "possible HeLa origin" of the Soviet cells.

"Of course, we must keep this information from becoming offensive to our Russian collaborators," he added. "Especially to Doctor Zhdanov." Victor Zhdanov was Moloney's counterpart, the senior scientist on the Russian side who had organized the presentation of the cell cultures.

Nelson-Rees felt honored by the request. But as the excitement of the phone call wore off, as he looked over his notes of the conversation, he also began to feel vexed. "Keep this from offending the Russians." It occurred to him that no one was genuinely interested in his work. As far as Nelson-Rees could tell, Moloney's deputies had for the past few months concerned themselves only with the awkward ironies and procedural infringements that it had stirred up. Now even Moloney's personal invitation seemed to him largely a warning that we must not insult our comrades in science.

Come on now, thought Nelson-Rees, this discovery resulted from a dandy piece of detective work. Not only that, it meant that HeLa contamination was still a problem—in fact it appeared to be a worldwide problem. What about a little acknowledgment? What about a little fanfare? What about my publishing it in the scientific press if you folks aren't interested? Or is the protection of Russian egos more important than spreading the word that Henrietta Lacks is still at large?

But in a letter to Moloney a few weeks later Nelson-Rees did not express himself quite so crassly. He outlined the material to be covered in his talk and added simply, "I would like to submit these results for publication as soon as you feel that the necessary amenities have been extended to all parties concerned."

On 12 November 1973, Nelson-Rees found himself in conference room number 10 at the headquarters of the National Cancer Institute in Bethesda, ten miles north of Washington, D.C. The classiest of the institute's meeting places, it was nicknamed "the Purple Palace" for the mauve carpeting and upholstery that adorned its outer lobby. A forty-foot table of heavy oak dominated the room. The table was oval and had a band of black leather along its edge upon which conferees could rest their elbows. There were microphones to amplify voices and earphones to offer simultaneous translations. The audio technicians and translators sat in a soundproof booth behind a darkened window set in one of the solid oak walls. The ceiling was made of wooden slats arranged in a waffle pattern. The carpet was thick and bluish gray. The chairs were black leather recliners with wheels.

On 12 November, for the meeting of the U.S.-U.S.S.R. Joint Subcommittee on Research in Oncologic Disease, the table was bedecked with miniature flags. Thumbelina versions of Old Glory and the Hammer and Sickle stood side by side, sticking up out of wooden hockey pucks in front of each namecard and water pitcher. On one side of the table sat the Americans, John Moloney and Nixon's appointed anticancer crusader Frank Rauscher at the center. On the other sat the Russians, arranged around Victor Zhdanov. Walter Nelson-Rees sat behind the American side of the table in a kind of raised gallery, the bleacher seats. About an hour into the meeting, Moloney looked over his shoulder and motioned for Nelson-Rees to come forward.

"We have examined the chromosomes and G6PD mobility patterns of six cell lines received in our laboratory in January 1973 from the Soviet Union via Bethesda," Nelson-Rees began. He was nervous. Like a kid putting a lit match to a pack of fire crackers, he was both looking forward to and dreading the explosions he planned to set off around him.

"Quinacrine-fluorescence indicated the absence from all cells of a Y chromosome." (Translation: We couldn't find a single Y chromosome in any of these cells. Although there are other explanations, this suggests all the cultures came from females.)

No reaction. He could see Victor Zhdanov staring at him: crew cut, mustache, and spectacles, but no expression on his face.

"G6PD mobility for all lines was of type A variant, characteristic of HeLa cells." (All six cultures contain the rare form of the G6PD enzyme, a form found only in the black population. Isn't that odd? Every cell seems to have come from a female who was black.)

Maybe someone coughed. Several of the American delegates may have nodded, their faces blank.

Nelson-Rees placed an enlarged black-and-white micrograph of chromosomes on the easel at the head of the table. Chromosomes labeled as coming from the Russian cells were arranged in vertical columns next to chromosomes from known HeLa cells. Barber pole patterns were clearly visible on all the chromosomes. "A trypsin-Giemsa technique for chromosome banding revealed marker chromosomes common to all cell lines, as well as to our culture of HeLa cells and to HeLa cells reported in literature." (As you can readily see, the fingerprints of the Russian chromosomes match those of the HeLa cells. The cells are identical.)

Dead silence.

It was as if he had been up there telling a long, complicated joke and had neglected to deliver the punch line. "Uh-huh, and then what happened?" several faces seemed to say.

There were others who refused to look at him. They riveted their eyes on the table before them, as if they were ashamed of something, as if they were hoping he would just finish the talk and spare them anymore of this crude, unspeakable subject.

There were also one or two faces fighting back the urge to smile at the whole humorous mess these Russian cells had been, but those faces were turned away for fear of breaching the solemnity of the occasion.

Nelson-Rees couldn't believe it. Maybe they simply failed to understand the data. Many of the Russian delegates were not well versed in the techniques of tissue culture. Yes, perhaps the significance just escaped them. Or maybe the Americans were still peeved that he had overstepped his bounds in checking out these cells and had risked unleashing a hoof-and-mouth plague on Detroit. Then again, maybe the whole lot of them had decided the most diplomatic way of handling his discovery was to ignore it.

Nelson-Rees had been girding himself for an explosion, and there wasn't even a burp. It was embarrassing. He felt suddenly silly up there at the easel in front of all these zombie-eyed scientists. Moloney thanked him for the presentation, and the meeting of the U.S.-U.S.S.R. Joint Subcommittee on Oncologic Disease was adjourned for lunch.

It was Victor Zhdanov who finally asked, "What does it mean?"

Zhdanov, Nelson-Rees, Moloney, and a few others had gathered after lunch in Moloney's office. It was quickly apparent to Nelson-Rees that the earlier session had been ceremonial. Now the Russians would have a more private and informal setting in which to discuss these curious findings. Under other circumstances Nelson-Rees might have been loudly indignant for being set up as part of the flag show in the Purple Palace. But he was so anxious to describe his work to anyone honestly interested that when Moloney asked him to repeat the highlights of his talk, he did so cheerfully. Then, when he had finished, Zhdanov asked his question.

"First of all," said Nelson-Rees, "it means you are not in control of the situation. You do not know what cells these are. You seem to have no control over where they come from or where they are going. But more importantly these cells are one and the same, and the viruses are one and the same, and it just isn't right to be calling them six separate isolates of human cells, each carrying a virus with potential human cancer value."

This time there was no stony stare. Zhdanov was visibly upset, though it is hard to say precisely what was racing through his head. He may well have been picturing the long list of Soviet research reports that were based on examinations of these cultures. In effect, Nelson-Rees was saying that most of the Russians' work on viral cancer in the last few years was questionable at best. Whatever he was thinking, Zhdanov quickly sought to take some advantage out of the catastrophe. "If you know so much about cells and can do so good," said the Russian, "why don't you come to Moscow? Show our people. Talk about what techniques you have."

"I'd be delighted," shot back Nelson-Rees, "if you promise I can get back out of the country."

The Russian laughed loudly.

Nelson-Rees joined him.

Suddenly it was one big happy meeting. Nelson-Rees jostled and joked further with Zhdanov. With some of the other Soviet delegates, he discussed opera and ballet, including the work of Rudolf Nureyev, a Soviet defector. If a few of the American bureaucrats were still sore about Nelson-Rees's independence in the handling of the Russian cultures—and miffed anew at his accepting this invitation to Moscow without consultation—they didn't let on. Finally, thought Nelson-Rees, the world is regaining its sanity.

The Journal of Virology soon proved him wrong. Now that the "necessary amenities" had been extended to "all parties involved," Nelson-Rees was set to publish his work. The friendly conclusion of the Bethesda conference had made him optimistic; the joint committee had gone so far as to formally agree it was "extremely important" that the findings be published. He prepared a manuscript and submitted it to the journal, which had just printed Wade Parks's conclusion that the Russian cells contained only monkey viruses. Certainly the journal would want its readers to know that the cells themselves were not what the Russians had claimed either.

The people at the journal didn't see it that way. "First of all it seems to be a gratuitous attack on the Russians," wrote one of the two technical judges of Nelson-Rees's report. "Secondly (and more importantly) it really is a footnote to history and not a study with new scientific implications . . . . I would hope . . . someday it would appear as a footnote." The other judge also suggested that the finding be presented only as a short blurb, and he preferred that it be presented in somebody else's journal.

Nelson-Rees was stunned and furious. Were these people—these people who are working scientists devoted to communication among colleagues—were they as worried as the bureaucrats about the politics of unpopular findings? It was the same incredible blindness. They couldn't see what was so obvious to him: the scientific community must be warned. The word must be spread that Gartler's list seven years ago was not the end of the story and that the cells of Henrietta Lacks are still on the loose!

When Parks heard the news, he wrote to the journal's editor, gently recommending that he publish some form of Nelson-Rees's paper. The editor sent a note to Nelson-Rees saying that having thought about it further he "would not be averse" to publishing a much shorter version of the report.

He would not be averse.

Nelson-Rees was not the sort of fellow to tell anyone to roll up a manuscript and stuff it, but he came close to making the suggestion to the editor of *The Journal of Virology.* Seething, he withdrew the paper and sent it to *The Journal of the National Cancer Institute.* The judges there were not as hostile, though several had trouble grasping the importance of the paper. It was finally accepted in June of 1974 and appeared in print three months later, almost two years after the Russian cells arrived in the U. S. and a year after the meeting at the Purple Palace.

The whole experience left Nelson-Rees sour. He had been wrong to think that anyone wanted to hear this bad news, except, ironically, the Russians. The funny thing is that the entire frustrating misadventure tempted him to jump in with both feet. It was almost enough to start him off on a personal crusade. Almost.

6

Keeper of the Cells

Growing up in Germany in the 1940s, Walter Rees was an outsider. He had been born in Havana and lived there until his parents separated when he was nine years old. Walter's mother then took him and his older brother to Karlsruhe, a town on the Rhine River. Living in Germany was the only way they could claim a small inheritance, left to his father, that would finance the boys' schooling. Walter's father, a German who sold farm equipment to sugarcane growers and later owned the Pepsi-Cola bottling plant in Havana, decided to remain in Cuba.

Being the sons of a German man, Walter and his brother were considered by the Third Reich to be German themselves. The term was *Auslandsdeutscher*, an outland German. It was an appropriate description for a young German boy who spoke only Spanish and English.

Nine months after Walter and his brother began classes at a strict, religious boarding school in southwestern Germany, Hitler invaded Poland. The Rees boys, like all the other children, joined the *Jungvolk*, the cub scouts of the Hitler Youth. Young Walter didn't enjoy the marching and the patriotic singing the way many of his classmates did. When he heard about plans for a military camp out in the country that summer, he announced that he wasn't going. An older boy, one of the group leaders who held the title of *Führer*, beat him severely for presuming he had any choice in the matter.

Walter was a good student. He learned to be disciplined and neat. He could do anything they asked of him; he just didn't feel a part of it. And he was happiest when he escaped. Whenever he could, Walter ran off to a section of the Black Forest that surrounded his school. There in the woods he and a friend built tree houses and pretended they were exploring the dark continent. Sometimes Walter would tell stories out of adventure books he had read or describe the real-life dreamland of Cuba.

No one spat at Walter as they spat at the little Jewish boys who were forced to wear Stars of David, and he was thankful for that. But as the war intensified and the Americans entered to fight Germany, he felt a prejudice against him and his family just the same, a sense that everyone knew they didn't belong. His mother, having been born in Havana during the U.S. intervention that followed the Spanish-American war, was considered to be an American citizen. Once when she was travelling to Messkirch, an official swept through the train, demanding to see identification papers. That was not unusual in wartime Germany. When he arrived at her compartment, however, the man smiled and said, "That's all right, you needn't show us anything. We knew you were riding with us, Frau Rees." They had been watching her.

Walter's mother taught him to be polite, helped him discover art and music, and encouraged him to be very good at whatever he did. She never said it quite this way, but what Walter heard was: be perfect. Show them that even if they won't accept you, you can be just as good as they are. Years later when Walter was in college, his mother tried to kill herself with sleeping pills. He thought that somehow he had disappointed her. "That led me to want to excel at many things," he told a friend.

"I wanted to do everything at least as well as the best around."

Walter was sixteen in 1945 when the American forces arrived to occupy Germany, though to him it seemed more that they were liberating the country. After years of rationing, suddenly there were chocolates and cigarettes and fresh fruit. What's more, there was no school. Walter got a job with the 485th U.S. Army Medical Collecting Company, a kind of mini-hospital that had set up near his home in Karlsruhe. He translated orders, typed letters, and washed and sterilized medical equipment. Although he had never been particularly interested in science, he decided then to become a doctor. After the war Walter returned to Havana, finished high school, and left for Emory University in Atlanta.

He dropped his medical ambitions after fainting in a dentist's office while a friend was having a tooth pulled—the sight of the blood had been too much for him—but he was already hooked on the study of plants, insects, and cells. He would sit at a microscope mesmerized as the cells of various animals doubled and redoubled before his eyes. Grasshopper cells put on the best show because they had the biggest and most visible chromosomes. That was what he fancied most, the chromosomes, those mysterious strands of genes that threaded themselves into intricate knots, emerged in orderly pairs of original and perfect duplicate, fell into formation across the center of the cell, and then marched to opposite ends, cueing the cell to split in two. In the spring of 1951, a few months after Henrietta Lacks's cervical tumor cells were put into culture, Walter Rees graduated with a bachelor's degree in biology. A year later, having earned a master's in cytology, also from Emory, he joined the U.S. Army. He had decided to remain in the United States, and the only way to quickly become a citizen was to enlist.

After three years at Dugway Proving Ground in Utah, working in the Army's biological weapons program, he received his citizenship, changed his surname to include his mother's family name, Nelson, and headed west, enrolling in a doctoral program in genetics at the University of California at Berkeley. There, under the guidance of Spencer Brown, a world renowned researcher who later became president of the International Genetics Federation, Nelson-Rees turned his idle fascination with chromosomes into serious study. Brown, known to some as Mr. Chromosome, taught Nelson-Rees the importance of finding and memorizing the chromosome patterns in certain cells, especially the oddball patterns. Look for the oddballs, said Brown, they're the clues to something interesting and useful.

Brown was an impatient man with fanatically high standards for laboratory technique. As exacting as Nelson-Rees had always been about his work, he found a new level of perfection in Brown's style, and he knew he'd never last if he didn't assume the same zealous approach. He found the challenge exhilirating, and he excelled. He became the favored member of Brown's retinue of graduate students. He actually looked forward to exams so that he could show his stuff. He began to talk of the beauty and elegance of a well-prepared slide of chromosomes. And he became known for leaving parties at ten o'clock to check a fly colony, count mealy bugs, or do some other chore in the lab, where he would end up spending most of the night. It was no hardship for him to be so devoted to his experiments. The truth was he felt there were better things to do than to dull his brain cells with alcohol and small talk.

When he had completed his doctorate in 1960, the chairman of the cell culture department at Berkeley's School of Public Health invited him to join a new lab they were setting up in Oakland to support the national cell bank, the one at the American Type Culture Collection in Washington, D.C. At first he applied his skills to the identification of animal cells by their chromosomes. In the late 1960s, when "The War on Cancer" heated up and officials at the National Cancer Institute wanted to rebuild the wreckage that Stan Gartler had left behind, they asked the Oakland lab to initiate and collect new human cell cultures to be used in the institute's most ambitious endeavor, the viral cancer program. That was where the money was. That was where the action was. And there was Nelson-Rees, quietly working his way up to the top of the new Oakland cell bank.

By the early 1970s, the institute was relying more heavily on the Oakland lab, and on its new director of cell production, Nelson-Rees, than on the American Type Culture Collection. If his people didn't have what the nation's top cancer researchers asked for, they went out and got it. Thymus one day, spleen the next, and tumor tissue from Asian and American Indian patients the week after. These institute scientists also depended on Nelson-Rees to check the identities of cultures they obtained from other labs. Those familiar with the Oakland operation knew it was Nelson-Rees's meticulous manner that made it so successful, so reliable.

"Walter's one major attribute," the institute's Jim Duff once said, "is perfection."

Wade Parks, the virologist who helped analyze the Russian cells, put it this way: "Walter's a little blunt sometimes. But he's never wrong."

There was, however, a joke about the status of projects such as the Oakland lab, projects that were funded by the institute but were not conducted by institute personnel as part of the in-house program. They were called the outhouse programs. Nelson-Rees used to repeat the joke goodnaturedly, though he did feel a distance between himself and the project leaders. There was no doubt that they knew who he was and appreciated his work, but in many ways he was still an *Auslandsdeutscher*.

7

Mug Shots

Officially they called it the Cell Culture Laboratory of the University of California at Berkeley, but in fact it was located in Oakland. Not in Oakland's fashionable residential hills but in the western flatlands, the marine industrial district. This was the part of town you saw first as you drove from San Francisco across the Bay Bridge: the storage tanks, the warehouses, the big Rustoleum red containerships at the docks, and, looming over it all, the gigantic metal cranes like a herd of mechanical dinosaurs foraging among the cargo at the water's edge. You knew you were in Oakland's harbor district when the roadside billboards stopped trying to sell you Johnny Walker Red and started pushing Tasco industrial valves.

The cell culture lab was housed in a former mess hall within a three-square-mile supply compound run by the U.S. Navy. There were no signs of Berkeley's rolling green lawns, the quads where students gather to picnic and play music, or the majestic marble buildings of higher learning. Here the office buildings were gray with dirty windows. Huge anchors and propellers were strewn around aluminum warehouses. In the alleyways stood dingy yellow forklifts. Scores of cargo containers, each the size of a Greyhound bus, were stacked high behind chainlink fences topped with barbed wire. Down the road was an outfit that hauled these "piggybacked" containers by rail and truck. It was called the Southern Pacific Golden Pig Service.

And yet inside the white building that held the cell culture lab there was no hint of the surrounding clutter. There was order and, through a certain set of double doors, there was cleanliness beyond imagination.

You couldn't go through the doors—they were locked—but you could peer through their windows into a bright, gleaming hallway. To get to that hallway you had to step into what looked like a large closet next to the double doors. Inside, a wooden bench ran from one side of the closet to the other, blocking your way. Where you entered, the floor looked like most institutional floors; its pattern of charcoal gray swirl was dulled by a hazy buildup. But just past the bench, it was a polished gem, the proverbial "floor you can eat off." On the wall was a large black-and-white sign:

Observe Change Rule
Never Come into Laboratory with Street Shoes Alone

The sign hung next to a series of shelves with forty-eight cubby holes, each labeled with the name of a lab worker, each holding a pair of shoes. Staffers had to change out of their street shoes and into a pair they wore only in the lab. For visitors there were white nylon booties in small, medium, and large.

The trick was to sit on the bench, put on a clean shoe (or slip a bootie over a dirty one) and swing it over to the clean side; then repeat the process for the other foot. Lab workers found the maneuver quite natural. But visitors often had trouble, accidentally bringing a clean bootie down onto the dirty side, in which case a new bootie was called for; or landing a dirty shoe on the clean side, which required a quick swabbing of the deck; or nearly falling off the bench in an attempt to avoid doing either.

The door in the clean section of the closet led into the gleaming corridor. Several individual lab stations extended along the hallway's left side. At the near end was the assay area, a large open space where newly arrived cells were logged in, examined without opening the flasks or dishes that held them, and stored until they could be analyzed more thoroughly and in the security of one of the "clean rooms" a little farther down the hall.

Each clean room was really a room within a room. The anteroom, about the size of three telephone booths, was where two technicians in surgical masks, caps, gowns, and gloves prepared for the task at hand, readying equipment, squirting alcohol over bottles of nutrient solution to kill bacteria, removing sterile wrapping on the glassware they planned to use. They then entered the pristine inner sanctum, closed the sliding door, and set to work: perhaps feeding a cell culture, transplanting portions of it into other flasks, or replacing the liquid medium with a fresh supply—changing its diapers, as the crew described it. The inner room had its own source of filtered air that kept the pressure there higher than in the rest of the lab, preventing airborne undesirables from drifting inside. When the door was opened, the air always rushed out of the room, never in. The technicians worked in pairs because it eased the load of monotonous chores. Open the flask, withdraw 10 ccs of medium, close the flask, label a new flask, open it, dispense 10 ccs of medium . . . . Working in pairs they could also keep an eye on each other. They were not to talk unless absolutely necessary; in spite of masks, talking increased the chance of spitting out microorganisms that might endanger the cell cultures. To eliminate the risk of cross-contamination, they worked on one cell line at a time, re-sterilizing the whole show before another line was brought in. It was an expensive, time-consuming procedure. They stayed in there for two to three hours at a time. But then obsessive care was what made Oakland so special.

When work on a particular culture was finished, any materials to be discarded were placed in pans full of disinfectant and taken to the far end of the hallway to the autoclave, a little chamber of hell that subjected its contents to lethal temperatures and pressures. All glassware was sent to be washed and rinsed in scalding, deionized water. Every working surface in the clean room was then wiped down with alcohol. Just in case some scrap of life remained at large in the room, and to kill off anything that might creep in later on, one of the technicians flipped a switch on the way out, bathing both compartments in ultraviolet light. On the door to each clean room was a sign that said in red letters:

KEEP CLOSED
Admittance to authorized personnel only.
Visitors and personnel not assigned to this area
contact W. Nelson-Rees.

On the other side of the corridor were the cramped quarters where Nelson-Rees and his co-workers mulled over the results of the lab work. Bob Flandermeyer had the office next to Nelson-Rees. He did most of his mulling while hunched over a table, pawing through a pile of photographs like a man searching for the right piece in a jigsaw puzzle.

Chromosomes, chromosomes, you could always find Bob Flandermeyer at that table sorting through hundreds of black-and-white enlargements of chromosomes. Long, straight chromosomes; short, stubby chromosomes; chromosomes that looked like bow ties, black ants, and licorice twists; and all wearing stripes, the irregular barber pole stripes known as banding patterns. Slowly, deliberately, *ploddingly*, Bob Flandermeyer would cast his big soft eyes over each photograph, studying the shapes and patterns.

Flandermeyer was a bear of a man with blond hair and a wide, friendly face. He spoke softly and a little ponderously. His laugh, loud and sharp, was startling punctuation to his sleepy speech. In early 1973, he left his job as a technician in a neighboring Navy laboratory. He had been working on a study of how bacterial diseases spread among troops, which had little to do with cell biology. Nevertheless, Nelson-Rees hired him to help apply the technique of chromosome banding to the identification of cells. Two other lab assistants before Flandermeyer had tried and failed to make it a practical system. The work was too tedious for his predecessors, but Flandermeyer mastered it.

The first part of the technique was nothing more than what karyologists, chromosome analysts, had been doing for years. The idea is to catch the cells in a stage of their growth cycle called metaphase, when the chromosomes line up across the center of each cell in orderly rows. You do that with colchicine, a drug derived from crocuses. Colchicine prevents the formation of tiny fibers that normally attach to the chromosomes and drag them to opposite ends as a cell prepares to split. It halts the chromosomes in their lineup like dancers in midstep. Then you add a solution to make the cells swell, which spreads the chromosomes apart so they don't overlap, and then an alcohol "fixative" to kill the cells. The next step is to attach a drop full of cells onto a microscope slide and flatten it into a layer thin enough that everything will be in focus for the camera. One popular method of flattening was the five-foot drop: holding the dropper at eyelevel, you aim at a group of slides arranged around your feet. From that height, a drop spreads itself into a broad, thin film, and if you're lucky the film ends up covering a slide instead of your shoes or the linoleum. For those with more flamboyant tastes, there was the crêpe suzette technique: a match is put to the slide, igniting the alcohol fixative, which burns away in a few seconds, dramatically reducing the volume of liquid in the drop. Once the sample is properly flattened, you mix in a little Giemsa stain and place the slide under a microscope. The chromosomes show up in silhouette, dark blobs without detail. Based on their shapes, a good karyologist can tell the species of the donor: man, mouse, or whatever. By studying the shapes and counting the number of chromosomes, he can also tell if the cells are abnormal. Victims of Down's Syndrome, for instance, have one more than the usual human complement of forty-six chromosomes.

That was conventional karyology.

In the new banding technique, before staining the cells, you dip a slide full of them into trypsin, an enzyme that eats away parts of the protein coat over the chromosomes. Then, instead of painting the entire outer coating of the chromosome, the Giemsa stain is absorbed into it, but only in certain spots, creating discrete bands of darkness next to blank areas. The problem is that the trypsin and the stain have to be at particular concentrations and temperatures, and the slides have to be dunked into each solution for precise periods of time. Even if you figure out the proper combination, the technique doesn't work unless you catch the cells at just the right moment in the metaphase stage. For some reason, the enzyme can't eat through the chromosome's coating otherwise.

Flandermeyer studied several days with a group of researchers who were using the technique at the University of Indiana. He returned to Oakland and began experimenting. Time and again he would prepare a bunch of slides, dip them in trypsin while counting off the seconds, then dunk them in the stain, only to find the familiar dark blobs where he had hoped to see barber poles. That meant he had exposed the cells to the enzyme for too long, allowing the stain to seep in everywhere. Or had he got the temperature wrong? Other times, when he didn't keep them in long enough or failed to catch the cells in metaphase, no readable patterns appeared.

Finally he hit on a reliable recipe for producing good, clear bands. He stained stacks and stacks of slides, then sat down at the microscope and scanned one after another, searching for the most photogenic cells. He photographed hundreds of them, printed up enlargements, cut out each banded chromosome, and mounted them all on white cardboard, grouped according to their banding patterns. After that the real work began.

First he had to memorize the patterns of normal human chromosomes. Every healthy cell has two copies of chromosome numbers 1 through 22, one from each parent, plus either a pair of Xs in the case of females or an X and a Y for males. So there are twenty-four different types of chromosomes, each type absorbs the stain in a characteristic set of bands—its mug shot—and Flandermeyer had to know each one on sight. He stared at the photographs for weeks. Through brute force of will he developed an eye for all twenty-four. But because these were normal chromosomes, present in every human cell, they were really no help in telling cell lines apart—not directly. He needed to know the normal chromosomes only so that he could recognize the misfits, the ones with funny shapes and weird banding patterns.

The misfits, the theory went, resulted from random errors in the duplicating process. Such errors occur frequently in tumor cells, where the usual mechanisms of replication and growth run amok. The lower arm of a number 5 chromosome might break off and attach itself to a severed upper arm of a number 3, for example. This unlikely goat with a lion's head, passed on to daughter cells, becomes a unique marker of that particular tumor and of any cell line derived from it. The odds were slim that an identical misfit would arise by chance in another cell line. And it was close to impossible for one cell line to randomly create a set of three or four misfits that matched those in another culture. Only cells with a common origin would have the same aberrant marker chromosomes.

For Nelson-Rees, who had always been partial to chromosomes, and who had learned the value of misfits from Mr. Chromosome himself, these markers were among the more powerful means of identifying human cell lines. When the National Cancer Institute made the

Oakland lab its hot new repository for human cultures, he decided that his quality control arsenal had to include a method of recognizing markers. But conventional staining wouldn't do it. To spot a marker you have to be able to make out which pieces of normal chromosomes had combined to form it. For that level of detail, you need to see the bands. That was why Nelson-Rees needed Bob Flandermeyer. And that was why Bob Flandermeyer, having learned every normal banding pattern, began staring at photographs of the abnormal markers, forcing his mind to associate each strange patchwork pattern with the cell line it had come from. The first markers he learned to spot were the four that other researchers had found in HeLa cells.

My God, he's slow, thought Nelson-Rees as he watched Flandermeyer pick his way through the photographs of unidentified cell cultures, looking for markers he had already memorized and memorizing the ones he came upon for the first time. "Ein Stier," Nelson-Rees had once called him, "an ox."

There was considerable art to the method, though. Flandermeyer knew that a banded chromosome's appearance often depended upon the angle from which he photographed it and whether it happened to be drooping or twisted or contorted in some other way. There was a lot of visual judgment involved, and Flandermeyer was not one to jump to conclusions.

Nelson-Rees and some of the other researchers who had learned to read banded karyotypes with Flandermeyer's help would often ask him to confirm their own assessments. He was, after all, the expert. On these occasions, Flandermeyer would stare at the photographs endlessly as the rest of them paced about. Finally, when he had convinced himself that yes, here was a marker chromosome in a suspect cell line, a marker that he recognized—but only when he was dead sure—he would smile broadly and say, "Well, lookie here."

With those words, Flandermeyer had helped Nelson-

Rees finger the Russian cell cultures as HeLa contaminants in the fall of 1973. Like a dogged assistant detective, he had pored over mug shots of the Russian chromosomes for hours until—"Well, lookie here"—he found all four of HeLa's known marker chromosomes. The Russian case had been their first practical application of the banding technique. Now, a few months later, Flandermeyer sat in his office, methodically sorting through the mug shots of two more cell lines of questionable identity. One was a line of breast cancer cells called HBT3, HBT for human breast tumor; the other was a culture of human embryonic kidney cells designated HEK.

Nelson-Rees had become suspicious of the two cultures several weeks earlier after walking in on a conversation between Adeline Hackett and Trudy Buehring, two researchers at the cell culture lab who worked independently of his cell bank. Hackett and Buehring were talking about this puzzling culture they had come across while studying the common structures of breast cancer cells. Spread out on the counter in front of them were photographs of cells from a number of established breast cultures. The pictures, taken through an electron microscope, showed that one of the cell lines, HBT3, had none of the features found in the three or four others.

"They just don't look like breast cells," Buehring was saying. "They don't look like *normal* breast cells, and they don't look like *cancerous* breast cells."

"Isn't that strange?" said Nelson-Rees. Then he remembered that he too had seen something funny in HBT3, almost a year earlier, when a California researcher under contract to the institute sent him a sample for analysis and deposit in the bank. Flandermeyer had not yet joined the staff and the only means of checking chromosomes was the conventional staining. Nelson-Rees's examination confirmed that HBT3 cells were of human origin, and that was as far as it had gone.

Except that he had seen an oddly shaped chromosome. His conventional staining had shown no detail, but the outline of the thing clearly resembled two little ears. Someone had called them Mickey Mouse ears. And come to think of it, he had seen a somewhat similar set of mouse ears in another cell line three years before that, in 1970. They were in a line of embryonic kidney cells called HEK. Maybe this oddball cell of Hackett and Buehring's doesn't look anything like a breast cell because it is actually from a kidney, thought Nelson-Rees.

He hurried off to pull the old sample of HEK out of the deep freeze and get Flandermeyer working on a comparison of the two cultures. The banding technique would show for certain whether the ears in one were the same as the ears in the other.

"Well," said Flandermeyer after an eternity, "lookie here."

The banding patterns of the mouse-eared chromosome in HBT3 precisely matched the patterns of the one in HEK. But that wasn't the end of it. There was another abnormal chromosome in HBT3, a new marker that Flandermeyer had never seen before, a long and boldly striped thing whose twin brother was smiling out at him from HEK as well. Two markers in common made an even stronger case that these were the same cell line. But . . . what's this? Something else? Uh-oh, lookie here: four more weirdos that Flandermeyer recognized instantly.

Nelson-Rees was amazed. Buehring and Hackett were flabbergasted. It appeared that HBT3 and HEK had been mixed up, as Nelson-Rees had suspected, but neither cell line retained its original identity. Both cultures were now HeLa, actually a previously unknown substrain with two new markers, but HeLa just the same. Instead of a cancerous breast culture and a normal kidney culture, what they had stumbled onto—and what many investigators interested in breast cancer and kidneys were no doubt wasting their time studying—were those familiar cervical cancer cells of Henrietta Lacks.

They were still reeling two weeks later when Flandermeyer emerged from a long session of mulling over mug shots to deliver more shocking news. Cell line HBT39B, yet another culture of supposed breast cancer cells that had been sent in for a routine check, also displayed the same six marker chromosomes.

It was like that point in every horror movie when the characters know there's no escape. True, there were no Saint Bernard-sized lumps of HeLa cells blocking the exits and gnawing through the telephone wires, but this horror story was for real.

First the Russian cell lines and now this. Three cultures from American scientists, cultures effectively selected at random, all taken over by the runaway cells of "our lady friend," as Nelson-Rees had started calling Henrietta Lacks. It reminded him of a remark made a few years earlier by a group of Johns Hopkins researchers who were marveling at HeLa's tenacity. "HeLa," they had written, "if allowed to grow uninhibited under optimal cultural conditions, would have taken over the world by this time."

The director of some other cell bank might have thrown out every culture of HBT3, HBT39B, and HEK, and left it at that. But to Nelson-Rees it didn't make any sense to merely note the problems, keep the cell bank pure, and let the calamities unfold elsewhere. Being the perfectionist that he was, Nelson-Rees couldn't stand the thought of all that error and confusion, not to mention the time and money undoubtedly being squandered by HeLa's victims. Being the keeper of the cells for the institute, the man armed with the latest techniques of cell identification, he felt almost duty bound to sound the alarm, to track down the fugitive cells of Henrietta Lacks, and to set things straight.

This was the quiet beginning of the crusade. There was no formal declaration. Nelson-Rees never called the troops together to say, "The war is on," and map out strategy. What he did, quite simply, was throw the entire operation into high gear. He began by tracing the path of the first breast culture. HBT3 had come to Oakland from a researcher at the California State Health Department, who, Nelson-Rees discovered, got it from a scientist at the Centers for Disease Control in Atlanta, who got it from Robert Bassin, an institute scientist, who originated the cell line in his Bethesda laboratory. Nelson-Rees wrote to all three. "Fully realizing the embarrassment to the originators of these cell lines and/or to the investigators from whom I obtained them, I would be very pleased to discuss this matter in detail with you to get at the source of this contamination, if indeed this is what is shown," he wrote. "I welcome and in fact must insist on further analysis of these and 'related' cell lines . . . ." It was like a note from the school nurse informing the parents that little Darlene had VD, and it drew the kind of reaction you'd expect.

Bassin was the first to respond. Being a careful scientist, he was well aware of the threat of HeLa contamination, which is why he had kept HeLa and all other human tumor cells out of his laboratory when he established HBT3 in 1972. Furthermore, his lab was housed in building 41, the institute's Emergency Virus Isolation Facility.

Building 41 was cut off from the world. No agent from the outside environment could leak in to jeopardize the purity of the experimental conditions there; none of the viruses or other nasty things they handled inside was able to leak out. There were no windows in building 41, no doors that opened without special clearance. The flow of air throughout the building was carefully controlled. According to institute legend, the high security so impressed a young medical intern who worked there one summer that it later moved him to write a science fiction novel about a deadly germ from outer space, *The Andromeda Strain*. In fact the book's author had never set foot inside building 41, but the facility's hermetically sealed atmosphere inspired those kinds of stories. It also made Bassin confident that Walter Nelson-Rees didn't know what he was talking about. Bassin telephoned to tell him so.

First of all, Bassin argued politely, it is very difficult to prove scientifically that two things are the same. It's simple to say they are different, of course. All you have to do is find some characteristics they don't have in common. But the fact that a few funny-looking chromosomes appear in both HBT3 and HeLa doesn't prove they are the same cell. Bassin questioned Nelson-Rees about this method of examining banded chromosomes. Could he be sure the markers matched identically?

Quite sure, said Nelson-Rees, adding that HBT3 was also carrying the A type of the G6PD enzyme, the type carried by HeLa cells.

That didn't prove a thing, Bassin came back, since the woman whose tumor established the HBT3 line was of northern Mediterranean extraction—Greek or Italian. Although it happened very rarely, type A had been known to show up in these populations. And even if the culture of HBT3 in Oakland truly was HeLa, wasn't it perfectly possible that the ones he worked with in Bethesda were bona fide? Maybe the California scientist who sent the cells in to be checked had contaminated them with HeLa in his own lab. Or maybe it happened in Atlanta.

Perfectly possible, Nelson-Rees agreed, which was why he needed to examine a culture from Bassin's personal supply.

The following day, a shipment of breast cancer cells took United Airlines flight 57 from Washington, D.C., to San Francisco. A messenger delivered them to Nelson-Rees, who passed them on to Flandermeyer for analysis. Nelson-Rees also shipped a sample to Ward Peterson in Detroit for the G6PD testing. As he waited for the results, he continued writing letters, making phone calls, notifying and debating researchers who were connected with the three contaminated cell lines. Like Bassin, none

of these researchers had been working with HeLa, or so
they claimed. None put much credence in the method of
chromosome banding. And those who conceded there
might be a problem unanimously pointed the finger elsewhere.

Ernest Plata, the man who had originated HBT39B,
the other breast culture, had considerably more in common with Bassin than any of the others. He had started
his cell line a few months after Bassin had got HBT3
growing. And he had done it just down the hall from
Bassin's lab, in the institute's windowless fortress.

Well, well, what a coincidence, thought Nelson-
Rees. It was obvious to him that a HeLa culture was running around building 41, masquerading as at least one
other cell line. And while the HeLa cells couldn't leak
out the carefully monitored vents or the air-locked doors,
they were far from trapped. Every so often Bassin, Plata,
or perhaps some other unwitting accomplice of Henrietta
Lacks would wrap up a few samples of contaminated
cells and mail them off. Nelson-Rees would have to have
the test results to be sure, of course, but it looked as
though the National Cancer Institute was distributing
HeLa cells, under various false names, all around the
country.

The last member of the triad, HEK, Nelson-Rees
could not trace back to its source. It had been established
ten years earlier by a laboratory that no longer existed.
Although it had been widely used in research, there were
no records of the donor's race, sex, or any other characteristics that could have been checked by his chromosomal
and biochemical techniques. The earliest cultures he
could find were in the hands of researchers at Pfizer Laboratories in Maywood, New Jersey, who had had them
since December 1964. He asked for a sample.

By February 1974, two months after Nelson-Rees
and colleagues had found the first indications of this
three-way contamination, they had called in and analyzed two or three specimens of each cell line. In every

one of them, Peterson had detected type A G6PD. In every one of them, including Bassin's personal supply of HBT3, Flandermeyer had found the marker chromosomes—the four traditional markers as well as Mickey Mouse and the one with the bold stripes, which they had named the Zebra.

Because this strain of HeLa had developed two markers not present in other known HeLa cells, Nelson-Rees and Flandermeyer decided that it must have been evolving on its own for some time, perhaps in an isolated environment such as an institute laboratory. Personally, Nelson-Rees suspected that this variant of HeLa had first contaminated HEK, the old-timer, and then gone on disguised as HEK to spoil HBT3 and HBT39B. But such reconstructions were a secondary concern at the moment.

It was time to get the word out. In the few months it had taken to track these three lines, they had come across two other popular cultures that were contaminated with HeLa—one a line of prostate cells, the other a culture of liposarcoma, a tumor of fatty tissue. There seemed to be HeLa contaminants everywhere they turned, though few people aside from the researchers Nelson-Rees had contacted had any hint of trouble. Nelson-Rees was still pushing, unsuccessfully, to publish the very first HeLa mix-up they had uncovered, the case of the Russian cells. Originally he had wanted to use the Russians' misfortune as a warning that HeLa might yet be alive and lurking around American laboratories as well. That warning was now well behind the times. These latest findings about five American cultures demanded some kind of all-out emergency alert.

"Anybody interested in working with characterized cell lines of bona fide purity of origin would be interested in this article," he wrote to Philip Abelson, editor of the journal *Science.* "Knowing that cell cultures presumably derived from human embryonic kidney, human breast carcinoma, human prostate tissue, and human liposarcoma cells are indeed derived from a human cervical carcinoma would certainly change the course of a number of research projects now in progress, and alter the interpretations in many publications already in existence involving these cells."

Science sent his manuscript to their technical reviewers, one of whom criticized Nelson-Rees's writing style, though he said he was sure that the findings were correct. The other said, "The main message of this paper is extremely important: that a surprisingly high proportion of cell lines are not what they are purported to be." In view of the rapidly increasing use of cell lines in research, the reviewer added, it is vital that Nelson-Rees's message be widely disseminated. From those comments, the editors at *Science* somehow decided that the report didn't quite meet publication standards. No thank you, they wrote back to Nelson-Rees.

How's that? Not interested in publishing the news that five cell cultures widely used in cancer research today are not what they're supposed to be? Was it happening again, just as it had with the Russian cells—this dead silence, this dumb stare, this gaping lack of interest in what he knew to be findings too incredible to ignore? Not this time. No, this time the sheer shocking momentum of his results couldn't be stopped. Rumors were already circulating. At the urging of several scientists and friends who had heard them, Nelson-Rees re-submitted the manuscript to *Science*.

It was then that Bob Bassin did a brave and unusual thing. After studying Nelson-Rees's banding data, and having performed a few tests of his own, Bassin conceded that his HBT3 cells might well be HeLa. This was no private confession. Bassin wrote to twenty researchers around the world to whom he had sent samples of his cell line, informing them of the bad news and asking that they send copies of his letter to anyone they had shared the cells with. Among other things, it was a graphic illustration of how far such an error might perpetuate itself.

Bassin also sent a copy of the letter to Nelson-Rees,

who forwarded it to the people at *Science* with a note saying, "You will, no doubt, appreciate the need for our publication." This time they appreciated it. The manuscript was approved and rushed into type. Nelson-Rees dictated a revised introduction and checked the galleys by telephone.

A few weeks later he was in Miami, attending the annual meeting of the Tissue Culture Association, when *Science* mailed out the issue carrying his report. A desperate fellow approached him at the pool of the Hotel Deauville, where the conference was being held. The man looked deeply troubled, as if he had just learned that tomorrow after breakfast the universe would blink out of existence.

"As I left the lab today," the man said blankly, "I saw the *Science* article. I just couldn't believe what I read."

Nelson-Rees had no idea how to respond. The lost soul turned and wandered off behind the lounge chairs and umbrellas.

Suddenly people were stopping him in the hallways to ask about chromosome banding and dropping by the dinner table to check a point about sterile procedures. The next day the place was positively buzzing about HeLa contamination, and the conference organizers asked Nelson-Rees to deliver an impromptu talk on his work. He didn't hesitate.

At last, he had got someone's attention.

8

Spreading the Word

The daily press knew how to handle this story. So what if *Science* buried Nelson-Rees's report in the back pages under the stodgy title "Banded Marker Chromosomes as Indicators of Intraspecies Cellular Contamination." The newspapers, properly horrified, played it on page one with headlines more to the point:

CANCER WAR SET BACK
GOOF COSTS 20 YEARS OF RESEARCH
A line of human tumor cells used by laboratories around the world for more than 20 years may have invalidated millions of dollars worth of cancer research, according to a scientist's report. . . . As a result, says the author, Dr. Walter A. Nelson-Rees, checks are in order for dozens of laboratories engaged in cancer research.—*Los Angeles Herald Examiner*

DEAD WOMAN'S CANCER CELLS SPREADING
Dr. Walter Nelson-Rees, one of the most experienced cell biologists in the world . . . has reported that many cell lines are by no means what they are thought to be by the laboratories handling them.—*Miami Herald*

A SHOCKER FOR SCIENTISTS

"The main situation has probably existed for years," said
the main author of the report, Walter A. Nelson-Rees,
a highly respected researcher. . . . Nelson-Rees said the
contaminating potential of the HeLa cells is well
known, but that sufficient precautions against it have apparently not been taken. — *San Francisco Chronicle*

All this publicity made no sense to a number of scientists. Why was Nelson-Rees taking bows now when
Stan Gartler had dropped the original bomb in 1966?

Part of the reason was that Nelson-Rees's paper was
printed in *Science,* one of the few technical journals that
nonscientists, particularly reporters, find accessible. One
section, prepared by the journal's news staff, was actually
written in English, and in the 7 June 1974 issue, the section carried a story that translated Nelson-Rees's article
beautifully. "If Nelson-Rees is right," wrote Barbara Culliton, "a lot of people may have been spending a lot of
time and money on misguided research. If, for example,
you are studying the properties of human breast tumor
cells, hoping to find features that distinguish breast cells
from others, and are, all the while, dealing unknowingly
with cervical tumor cells, you've got a problem." That
was plain enough even for a newspaper reporter to understand, and to embellish and bang out for the morning
edition.

But what really made Nelson-Rees a media star was
the dramatic background of his shocking results: "The
War." In Gartler's day, HeLa contamination had been the
dirty little family secret of the tissue culture crowd. Its
broader impact was not obvious. In 1974, however, "The
War" had been officially declared and raging for several
years. Everybody knew that the nation's most brilliant
medical experts were at this very moment working feverishly against the scourge of cancer. It was a national pri-
ority.

Nelson-Rees's message made this large and serious
effort seem a little silly. Sure, the institute was spending

millions of dollars sending its brave recruits over the top against the enemy. But it turns out our boys were shooting with *blanks*! It was a scandal, and there's nothing the press likes better than a scandal. Besides, this story came with a bonus—the awkward adventure of the Russian HeLa cells, until then unpublicized. The reporters loved sprinkling that one in: we've not only screwed up cancer research, folks, we almost blew détente on account of these crazy cells.

Nelson-Rees returned from his week in Miami like the local boy come home from battle. Reporters were still pursuing him. Friends showered him with congratulatory phone calls and letters. And the laboratory's resident bard immortalized his accomplishments in a limerick that appeared on one of the office bulletin boards:

A perceptive young Nelson named Rees,
Dumbfounded genetic police
When HeLa he found
To abound all around,
In cell lines from West and from East.

In addition to all the excitement it generated, the publication of their report brought great relief to Nelson-Rees, Flandermeyer, and the rest of the crew. It had been frustrating to know what they knew without having a means of broadcasting it. Now that the news was out, it might be easier to spread the word about future screw-ups.

Yes, spreading the word, that was the goal here. The publicity was fun, of course. It was nice for an "outhouse project" to steal the show for a moment. But getting the news out to those who needed to hear it—the scientific community—that was the main point of all this. And the publication of their paper was really just the beginning.

Frantic scientists had been calling and writing from all over the United States and several other countries to

request copies of the article. In a few weeks Nelson-Rees's
secretary had mailed out all 400 reprints and had to order
another batch from *Science*. The author himself was in
equally high demand. Would he give a lecture at the
Stanford School of Medicine? Could he address a meeting of viral cancer researchers in Hershey, Pennsylvania?
Would he brief a group at the Argonne National Laboratory in Illinois?

Nelson-Rees hit the road, preaching and proselytiz-
ing like biology's Billy Graham. At one stop he would tell
the stories of such victims as Bassin and Plata who,
through lack of vigilance, fell prey to HeLa's sabotage. At
another he would describe the waste and futility of trying
to learn about breast or prostrate or kidney cells by studying cultures of cervical cancer. And he would always conclude with the exhortation: Never trade cells without reliable information about what they are and where they've
been. And always double-check them, before and after
your experiment.

In some respects Nelson-Rees's early evangelizing
looked like a touring revival of Gartler's performance at
the Bedford Springs Hotel. Like Gartler, Nelson-Rees was
telling audiences they had torpedoed years of their own
work by being sloppy and letting HeLa creep in on them.
He encountered the same reactions, shock and skepti-
cism. And he too was offering new tools, chromosome
banding along with other techniques, to help set things
straight.

But Nelson-Rees soon added a new message that was
less scientific in tone and more philosophical, or perhaps
more political. It was at a 1975 meeting of cell culturists
and cancer researchers in Lake Placid, New York, that he
began to talk about two different reasons for cell mix-
ups. One was simple sloppiness. "It can be combatted
in individual laboratories by adherence to increasingly
stricter techniques," he explained. The other effect was
"more lasting and insidious." It had to do with research-

ers' attitudes, "frailties of the human ego . . . exigencies of profit margins . . . the threat of cuts of support in contractual arrangements."

Most members of the audience had gone into his talk thinking they had a pretty good idea what Nelson-Rees was going to say. But when he hit this stuff about scientists' attitudes and frailties of the human ego, they didn't know quite what to make of it.

It was curious, he told his fellow researchers, how well they handled a problem involving bacteria, for example, or viruses. Most of them quickly faced up to it when a cell culture became spoiled by such infectious agents. Not so with cell mix-ups. They seemed to take it personally when someone claimed their cultures had been overtaken by other human cells, he said.

"Cases of cellular contamination have been known to precipitate lengthy diatribes and are the reason for lectures such as this one," he clucked. "This kind of contamination would certainly be easy to control if one could frankly and readily discuss it and eradicate it without fear of offending colleagues' feelings."

At the moment, said Nelson-Rees, there was no friendly forum for discussing cell mix-ups and no means of rapidly notifying the scientific community of contaminated cultures. The audience raised a few eyebrows in disapproval as he said, " I would now like to describe to you how difficult it was for us to publish our results on those cultures which we have vouched are HeLa derivatives."

He began with the story of the Russian cells. He recounted how reluctant institute officials had been to believe his results, how for months they merely ignored the rapidly accumulating data. He read aloud the reviewer's critique that accused him of making a "gratuitous attack" on the Russians, and he named a few names. Then he moved on to the American cells: HBT3, HBT39B, HEK, and the others. He explained that *Science* had originally turned down the now celebrated report, quoting sarcastically from Abelson's rejection letter: "The manuscript and the referees' comments are enclosed. I *trust* the comments will be *helpful* to you when you prepare the paper for submission *elsewhere.*"

Stan Gartler had certainly never done this. He had presented his findings, disturbing as they were, made a couple of suggestions, and taken his leave. But here was Nelson-Rees hauling out the dirty laundry. He was pointing out the stains and explaining how they had got there, and more than a few members of the audience were starting to squirm.

To say that a few cell lines had got shuffled around, that was one thing. But to suggest that scientists were letting their egos get in the way of good science or that they avoided publishing important information because it might be controversial . . . well, it just wasn't done.

Robert Stevenson was frequently in the audience at Nelson-Rees's presentations, though Stevenson never squirmed. Knowing Nelson-Rees as well as he did, Stevenson more or less expected him to say something startling.

Nelson-Rees and Stevenson first met in 1962. Nelson-Rees had just been hired as a research associate with U.C. Berkeley's School of Public Health. He was sent to Washington to tour the nation's central cell bank facilities at the American Type Culture Collection and to talk with Stevenson about the school's role in assisting the new program. As head of the National Cancer Institute's cell banking program, Stevenson was helping to organize and fund the network of outside support laboratories. These included Lewis Coriell's Institute for Medical Research in New Jersey; the Child Research Center of Michigan, where Cyril Stulberg and Ward Peterson were based; and Berkeley's new cell culture lab in Oakland.

Bob Stevenson was not a typical bureaucrat. For one thing, he spoke his mind. He routinely called those responsible for the confusion in cell culture rank amateurs, and he said it in print as well as in conversation. Until 1960 when the institute recruited him to develop high-quality virology research materials—cell lines, in particular—he had been head of a tissue culture lab at the Naval Medical Center in Bethesda. He had seen his share of accidentally contaminated cells, which gave him a first-hand appreciation of what the confusion meant. In fact, though still in his early thirties at the time, Stevenson was one of the old guard who discovered the original mix-ups among animal cells and called for a central cell bank in the late 1950s, years before Gartler's first discovery of HeLa contamination.

Stevenson had an honest, friendly manner that fit his cherubic face. His cousin, an illustrator for children's textbooks, had used that face as a model for Dick of the *Dick and Jane* elementary readers. But Stevenson also had a devilish grin and a mischievous streak. He liked to "stir up the muck," as he put it, "to get people thinking about stuff they don't usually think about, but should." When he left the lab bench for a desk at the institute, Stevenson thought of himself as being scientifically castrated. The only way to keep contributing, he figured, was through others, by encouraging them to do provocative experiments and to trumpet their own findings.

Maybe that was why he and Nelson-Rees hit it off. Nelson-Rees was a kindred spirit, a young and energetic perfectionist who was still in the lab. They quickly became allies, beginning a long partnership. Stevenson made Nelson-Rees a member of the advisory board to the cell bank, the group that judged whether a particular culture was qualified to be in the "reference library." Based on his analyses in Oakland, Nelson-Rees would tip off Stevenson about suspicious cell lines. As executive secretary of the board, Steveson would then ensure that the proper probing questions were asked at the meetings. Nelson-Rees would report, for instance, on the nonhuman-looking chromosomes of a purportedly human culture, and the culture would be reviewed and rejected.

Later, when Stevenson moved up through the hierarchy and then left the institute in 1967, passing oversight of the Oakland lab to Jim Duff, he continued to encourage Nelson-Rees. In telephone pep talks—usually when Nelson-Rees had uncovered something controversial as in the case of the Russian cells or the three American cultures that followed—Stevenson would ask him, "Is your work good?"

The answer never varied.

"Well, then," Stevenson would say, telling Nelson-Rees what he wanted to hear, "it doesn't matter that the shit is going to hit the fan."

After Nelson-Rees's report appeared in *Science*, Stevenson kept trying to get him more involved in activities of the Tissue Culture Association, the professional group for cell culturists. Stevenson, who by then was back at the institute as manager of the Frederick Cancer Research Center, formed the association's Committee of Standardization, Collection, and Distribution of Cells and Tissues, and appointed Nelson-Rees as a member. Together they struggled to convince the group of the need for a listing of bona fide cells to be circulated among the association's members. They also launched a campaign to persuade journal editors to require complete, authentic descriptions of every cell line used in a published research report, including a summary of what tests were used to verify that the cells' species, sex, chromosomes, enzymes, and other traits matched those of the purported donor. Nelson-Rees and Stevenson never tired of saying how ludicrous it was that the association's own journal, *In Vitro*, had no such requirement. As chairman of the group, Stevenson also had the committee officially endorse Nelson-Rees's findings and recommend that he keep up the good work. And being a quintessential organizer, Stevenson was constantly setting up conferences to stress the need for careful monitoring of cell lines, conferences that often featured a presentation by Nelson-Rees.

For two friends, both interested in stirring things up, it was funny how different their styles were. Stevenson was easy-going, informal, naturally likeable. He was fond of telling stories, such as the one about the New York medical examiner who performed rectal autopsies so that relatives of the deceased wouldn't notice the incisions. Nelson-Rees on the other hand was stiff, painfully honest, and sometimes holier-than-thou; he just automatically got certain people's dander up. If tact consists in knowing how far to go in going too far, as Cocteau once wrote, then Nelson-Rees was tactless.

Stevenson had tried to explain the benefits of a little diplomacy to Nelson-Rees. "You're very valuable, Walter," he once said to him. "You have a real talent for sniffing out trouble. But it might be advantageous to talk more softly and still carry that big stick."

"You cannot ask a leopard to lose his spots, Bob," Nelson-Rees would respond, "I am what I am."

And what he was looked very much like obnoxious to some people. He was so high and mighty about HeLa cells and so full of himself. In his talks he had started to make comments like, "This slide shows the *now well-known* marker chromosomes of HeLa. . . ." or, "In the case of the *now well-known* Russian HeLa calls. . . ." as if to say, "You've all heard of me, no doubt, and my very important and well publicized work."

He described his 1974 trip to Russia like a missionary returning from the jungle with word that the heathen had seen the light. Not only that, he said, they didn't have any of this trouble with ego or funding worries.

"Contrary to being an insult to our Russian colleagues, our data resulted in a most generous invitation to Moscow last November," he told the crowd at one meeting. "There I visited five major institutes and at least fifteen individual laboratories, gave three seminars (translated simultaneously), and witnessed the initiation of at least four new cytogenetic control sections for cell line monitoring at different institutes. As you can see, I was also allowed to return after a very warm and friendly reception."

All right, Walter, you're terrific! Could we move on to something interesting now? Some thought all this talk about sterile conditions and strict monitoring sounded more like a fifth-grade personal hygiene class than serious scientific discussion: never swap cells, boys and girls, and remember to floss every day. Whenever he rose at a conference to ask a question or make another one of his public service announcements for care and quality, these people would glance at each other and roll their eyes.

There were many, of course, who took his talks seriously. Even Duff and the other bureaucrats at the institute were convinced his work was important and his motives were unselfish, though he did make them nervous. But there were also a few who saw Nelson-Rees as a dangerous publicity hound, a Joseph McCarthy of the cell culture circuit who wanted to further his own career by ruining the careers of others. They bristled every time he challenged colleagues about the identity of cells used in experiments or asked how that identity was determined. And he was doing that a lot these days.

One afternoon in June of 1975, Walter Nelson-Rees and Relda Cailleau were standing in the lobby of Montreal's Queen Elizabeth Hotel screaming at each other. The occasion was the annual meeting of the Tissue Culture Association; it had been about a year since the publication of Nelson-Rees's report in *Science*.

Nelson-Rees had begun the conversation by asking about a cell line Cailleau had initiated sixteen years earlier, a lung culture called MAC-21. Cailleau exploded.

"You're an assassin. You're vindictive. You're just out to wreck people's reputations!"

"Relda, there's no need to scream at me," Nelson-Rees screamed back at her.

"We know all about cell culture. You have no right to attack us."

"Relda, the very way you're telling me to mind my own business is what *makes* it my business!" Nelson-Rees's Dudley Doo Right tenor rose and fell in pitch, sounding over the din in the lobby like an air raid siren.

They were quite a sight. Nelson-Rees, six-foot-three and gangly, towering over Cailleau, all of five feet, a round, tough-talking bundle of volatility. She had a large nose, darting eyes, and thick, dark eyebrows. As she argued in her high scratchy voice, she tossed her head from side to side and waved her arms around.

"Relda" was her mother's maiden name spelled backwards. The former Rose Adler, her mother adopted "Rose Relda" as a stage name while singing at the Opera Comique in Paris. "I can curse real well in French," Relda sometimes told new acquaintances to whom she was describing her background.

Walter and Relda went way back, though Relda went back even farther on her own. In the early 1960s, when the Oakland lab and Nelson-Rees were both just starting out, she was already an accomplished cell culturist working at the U.C. San Francisco Medical School. They would bump into each other all the time at local and national scientific gatherings, always as courteous colleagues, if not close friends. In 1966, however, they had a falling out over the question of an evening's entertainment—an incident that Nelson-Rees believed to be the start of a cold war between them.

The way Nelson-Rees told the story, he was chairman of the organizing committee and she was in charge of rooms and facilities for a San Francisco meeting of the Tissue Culture Association. Nelson-Rees had proposed a cruise on the bay with a buffet dinner and California wine tasting. Cailleau, sure that the summer fog would spoil the festivities, fought the idea but was overruled. The cruise, complete with a watercolor sunset and radiant full moon, was a rousing success. Although they lived less than a mile from each other in San Francisco, Nelson-Rees and Cailleau rarely spoke again. In 1970 she moved to Houston to work at the M.D. Anderson Hospital and Tumor Institute.

But of course it wasn't the cruise that was bothering Cailleau at the moment. In the past year she had watched Nelson-Rees crusading, making speeches, basking in the limelight, and she didn't like what she saw. His methods were antagonistic, and his goals appeared more destructive than constructive. On top of that she couldn't stand his attitude.

"He's so busy showing off what a great guy he is, how he's always right," she once complained. "He's so supercilious it riles the hell out of me."

And now he was trying to pull that stuff on *her*, trying to attack *her* cell line, MAC-21.

Nelson-Rees always insisted he wasn't attacking anyone, really. He said it was just part of his normal information-gathering routine. He scrutinized every cell line sent into Oakland for analysis, he scanned the journals for signs of mix-ups, and he walked into every conference with his eyes and ears wide open, constantly searching for clues, continually trying to make connections.

In the case of MAC-21, he had stumbled across its suspicious nature about a month earlier while checking the validity of two other cell lines developed by a colleague of Cailleau's at the Anderson Hospital. It happened at a meeting of cancer researchers in San Diego. Nelson-Rees noticed in the program an announcement of two new breast tumor lines, SH2 and SH3. Even in 1975 breast cells were very hard to cultivate. Researchers had put hundreds, perhaps thousands of tumor fragments into culture, hoping for a long-lived cell line to take root, but there were only one or two established cultures. Bassin's HBT3 and Plata's HBT39B had been celebrated additions to the short supply until Nelson-Rees came along. The report of two brand new breast cancer lines was therefore of great interest.

What made Nelson-Rees suspicious of SH2 and SH3 were a couple of characteristics described in the program. To begin with, both cultures were said to have come from Caucasian patients, yet both carried G6PD type A, the enzyme variant that is virtually nonexistent among Caucasians. Their early histories were strange too. SH3 had sat dormant, barely growing in the culture dish for thirteen months before it bloomed into a fast-expanding cell line; SH2 had taken three years to do the same. Nevertheless, the announcement took pains to point out that the chromosomes and several other chracteristics of the SH lines were clearly different from those of HeLa cells.

Nelson-Rees went to a talk given by a member of the Anderson group, patiently sat through it, and rose during the question-and-answer period. "Have you any information," he asked, "on whether chromosome banding techniques have been applied to these cells to exclude the possibility that SH2 and SH3 are, first of all, not one and the same cell line, and secondly that they do not have a whole complex of marker chromosomes in common with many now well-known HeLa strains?"

Gabriel Seman, the French-born virologist who had initiated the SH lines, answered that in fact banding had been done by collaborators at the Institute for Medical Research in New Jersey, who assured him that they were not HeLa cells.

One of the advantages of Nelson-Rees's public displays was that it turned up extra clues, often volunteered by colleagues who smelled something rotten but were not as ready as he to stand up and make a spectacle. The SH cells might have wriggled off Nelson-Rees's hook except that as the audience was filing out, Lewis Coriell, the director of the New Jersey institute, pulled Nelson-Rees aside to say that he too was bothered by the unusual characteristics of these cell lines. He said he would discuss them with his karyologist, Bob Miller, who had done the analysis for Seman.

Back at Oakland a couple of days later, Nelson-Rees received a letter from Coriell explaining that months earlier Miller had indeed concluded from his analysis of banded chromosomes that SH2 and SH3 were identical cell lines. Why Seman had reported them to be two independent lines was a mystery. In addition Miller had seen some similarities between the SH chromosomes and those of a HeLa culture he had in his lab, but he had decided the SHs were not HeLa because there were also many dissimilarities. "You have looked at more HeLa cell lines," wrote Coriell, "and he [Miller] would like to have your opinion of his interpretation." Mug shots of the cells were enclosed.

Three chromosomes in each SH line were so obviously HeLa markers that Flandermeyer "the Slow and Deliberate" made his pronouncement the very same day the letter arrived. The markers, coupled with the observation that these cultures from Caucasian women were both carrying the black type A enzyme, convinced Nelson-Rees that both cell lines had been taken over by HeLa. He notified Coriell and Miller.

"In view of this additional information it would seem probable that the lines are contaminants," Miller then wrote to Seman. "I realize this differs somewhat from my original interpretation but hope that the situation is now clarified."

Yes, there was *somewhat* of a difference between saying SH2 and SH3 are cultures of breast cancer and saying, "Oops, they're HeLa."

Seman was incensed, though it is hard to say what upset him more: Miller's new diagnosis that the precious breast cultures were Henrietta Lacks's well-traveled cervical tumor cells, or the fact that Miller dared send the SH photographs to another investigator. "Wizzout my permission!" as he put it. Seman was even angrier with Nelson-Rees for using information that Seman decided had been obtained unscientifically and illegally. "Eeet's a case for my attorney," he declared.

Seman's colleagues at Anderson, Relda Cailleau among them, were equally outraged. Nelson-Rees had acted irresponsibly and unethically, they all agreed. It was the same trick he had played on other unsuspecting researchers. To them the question of HeLa contamination was somehow a secondary concern.

Not to Nelson-Rees, though, who for the moment was oblivious to the reputation he was building at the Anderson Hospital. Having nailed SH2 and SH3, he turned his attention to MAC-21. He remembered seeing a reference to MAC-21 in the paper announcing the SH lines. Specifically, the report had said that the SH cells were like MAC-21 cells in that they had the type A form of G6PD.

Now MAC-21 had been around a long time, used extensively in lung cancer research, yet Nelson-Rees didn't recall ever hearing that it had this unusual trait. So it was that a month later when he saw Relda Cailleau descending the stairs at the Queen Elizabeth Hotel in Montreal he walked up and said, "What's this I read about MAC-21 being type A? Have you ever checked it?" That's when Cailleau erupted.

"You're doing this to everyone, are you? You're assassinating people!"

"Look, we know there's a problem with SH2 and SH3. They were both type A G6PD and we found out they have HeLa chromosomes."

"MAC-21 is not HeLa. I'm absolutely positive."

They went on for ten minutes like that. Finally Nelson-Rees went into one of his commercials. Cailleau might be interested to know that he and Bob Stevenson were firing up the cell standards committee of the association to advertise the need for careful controls and regular monitoring of cell identities, he said. "But I don't have to tell *you* the importance of keeping things straight. As you know, Relda, I'd be glad to check any of your cultures for you."

At that, Cailleau stomped out of the lobby.

It was during this early period of the crusade, within a year of the *Science* article, that an amusing telegram arrived at the Oakland lab. The unsigned message offered Nelson-Rees a high-paying and prestigious new job in research. It also offered a one-way ticket to get him to this very attractive position—in Uganda.

9

Damage Report

It was too bad that Nelson-Rees didn't have an accountant traveling along on the crusade, someone to assess the piles of spoiled research he had dug up and left rotting in his wake, a bookkeeper of the bereaved to record the number of hours squandered and research dollars frittered away. The newspapers said millions had been wasted by HeLa's unexpected spread, but there were no official damage reports. The bureaucrats at the National Cancer Institute never tried to survey the wreckage, and they probably would have failed had they attempted it. Very few of HeLa's victims, even those asked point-blank, would ever detail how much time, effort, and money they had wasted, or how many colleagues they had led astray.

Take Ernest Plata, for instance, the cultivator of HBT39B, one of the two HeLa-contaminated breast cultures that came out of the institute's building 41. Plata established HBT39B in the summer of 1971 and studied it on and off over the next two and a half years, hoping to find viral clues to the cause of breast cancer. After Nelson-Rees convinced him that his malignant breast cells were in fact malignant cells of the cervix, Plata put them permanently on ice and notified the six or seven people with whom he had shared them. No harm done, though, according to Plata.

"I wouldn't necessarily count it as a loss," he said many years later. "In some ways, it served to accelerate awareness and prevent many more losses."

Not every researcher who had a run-in with Henrietta Lacks considered it such a lucky break. Some simply refused to believe it. They continued working with their cultures, many of them honestly convinced the cells were bona fide in spite of Nelson-Rees's test results. Needless to say, they offered no damage estimates.

Nelson-Rees himself never cared much for figures. Besides, he was too busy tracking HeLa down, convincing people he was right, and lobbying for reform. "The enormity of the problem is obvious," he would say, pointing to his growing list of HeLa-contaminated cultures. "There's many a sad tale here."

One of the few sad tales that did have a price tag was the story of HBT3, the other breast cancer culture from building 41. The mix-up of HBT3 was no more scandalous than the mix-up of any other cell line. It's just that Bob Bassin and his group reacted pretty reasonably to the news about their culture, and they candidly discussed details of the mishap.

Like Plata and many others, Bassin and co-worker Brenda Gerwin had been looking for signs that a virus was the cause of breast cancer. First, they needed a living piece of breast cancer. After trying nearly 100 tumor fragments, Bassin finally got one to spawn a cell culture, which he christened HBT3. Although there were no viruses to be found in HBT3, Bassin and Gerwin hoped to turn up a clue that viruses had been in the tumor at an earlier time and had left their mark on the cells. What they hoped for was a telltale enzyme produced by the virus, an enzyme called reverse transcriptase.

It took Gerwin and a technician a year, devoting half

their days to the effort, to purify and isolate an enzyme from the cells and to identify it as reverse transcriptase. In August 1973, Gerwin, Bassin, and their associates published a report in the *Proceedings of the National Academy of Sciences* announcing the find.

Gerwin and her assistant then spent another three or four months combining the enzyme with blood serum drawn from patients with breast cancer. The theory was that if this enzyme were really manufactured by viruses that caused breast cancer, then there would be antibodies in the blood of breast cancer patients that would attack the enzyme. But they found no evidence of an attack. It was about that time, toward the start of 1974, that Nelson-Rees's letter arrived, making it clear why the blood of the breast cancer patients had not reacted to the enzyme from their "breast cancer" cell line.

It took another month or two, but Gerwin and Bassin became convinced that HBT3 was actually HeLa, and that they had been wasting their time. Had it been another kind of cancer cell that slipped in, they might have salvaged a few observations about that cancer's relationship to viral enzymes. But not in the case of the HeLa line. HeLa had been in so many labs and so many different microbiological environments, they knew, that it could have picked up all kinds of things that had nothing to do with viral cancer. They abandoned their work on HBT3.

Nearly a decade later, Gerwin and Bassin calculated the damage. Gerwin figured the half year of purifying and identifying the enzyme was not a total waste since some of the techniques developed she applied to other experiments. Perhaps three of those six months were for naught. She and her technician wasted another three months testing the patients' serum against the enzyme. The combined six-month salary of a Ph.D. and a technician in the mid-1970s was about $20,000. To that Gerwin added support costs, anything from the plumbing and heating to cell culture materials, which she guessed came to another $20,000. The overall loss for Gerwin's end of the project, then, was about $40,000.

Before Gerwin even started her work, however, Bassin had spent three months nursing the cells into a sustainable culture and performing various tests. He thought it cost at least $25,000 in wages and laboratory expenses. In other words, the HeLa cells that overran HBT3 inflicted $65,000 worth of damage on the group that initiated the culture.

The damage didn't end there, of course. Bassin had sent twenty-seven shipments of what he thought was HBT3 to twenty research teams around the world, from the University of Rhode Island to the Karolinska Institute in Stockholm. The shipments went out between August 1972 and August 1973, meaning that scientists could have been working with the cells for anywhere from eight months to a year and eight months before they received Bassin's notice about the HeLa contamination. If only four of the twenty groups invested as much effort in this culture as did Bassin's team, the total damage would be $260,000.

There were only two other sad tales for which anyone would try to figure a tab. One was the Russian caper, Nelson-Rees's first brush with HeLa. According to Wade Parks, leader of the team of American virologists who analyzed the viruses in the six Russian cultures— the viruses some hoped would include *the* human tumor virus, the ones that turned out to be monkey viruses —the laboratory work alone cost $50,000 to $60,000. As for the total cost—including such incidentals as investigators' time, the price of hiring a commercial lab to grow large quantities of the cells, and the expense of elaborate meetings to discuss the findings—Parks estimated it many years later at a quarter-million dollars.

And then there was MA160, a bogus culture of human prostate cells that Nelson-Rees featured in his 1974 *Science* article. Seeing that article, a cell biologist at the

University of Colorado named Mukta Webber scanned the scientific literature and found that six recently published studies had been based on the erroneous assumption that MA160 was a prostate cell. At the time she gave no quantitative assessment of the harm done, though researchers familiar with the field would eventually estimate the price of those six misguided studies at $30,000 to $40,000 apiece—a total of $210,000 to $280,000. As it turned out, Webber had found only a portion of MA160's misadventures.

But taking even the minimum damage estimate of $210,000 for MA160 and adding a quarter-million each for HBT3 and the Russian cells, the cost of just three instances of HeLa contamination approached $1 million. To say nothing of the two other cultures on Nelson-Rees's 1974 list or the monetary equivalent of fifteen years of work lost in the early era of cancer research, long before Nelson-Rees joined the chase. Or the waste due to holdover cultures from those days, lines such as the Chang liver and the WISH amnion, which, despite being branded as HeLa, were still surfacing periodically in published research reports as bona fide liver and amnion cells. Or, finally, the many new HeLa-contaminated cell lines that Nelson-Rees and Flandermeyer were still uncovering.

Bob Stevenson, at one time the man in charge of cell cultures for the institute, was one of the few officials ever to guess at the total damage HeLa had done. Some ten years after the newspapers first reported the story, Stevenson would look back and figure they were right. By the time Nelson-Rees had published his report in *Science*, the losses added up to millions of dollars.

"Millions is probably the right ballpark," Stevenson would say. "But you're not going to get anyone to admit that. It's like something uttered in a confessional."

10

Provenance

Walter Nelson-Rees is driving away from Oakland, toward the East Bay hills and Orinda. Although the passenger has requested no tour, he is from out of town and Nelson-Rees cannot resist. "This street we're on was the old Broadway of Oakland. That's old Tunnel Road over there. The freeway on our left was built with all the fill they dug out of the hill up ahead to make the Caldecott Tunnel. On the right here is a new sports complex, a recreation area I guess you'd call it, that is in the process of being constructed. . . . "

He turns the car back in the direction of Oakland and heads downtown.

"This church over here on Castro Street used to sit one block that way on Brush Street. They actually lifted the entire thing and moved it for the freeway."

The visitor begins to speak, but—"Originally, I think it was Greek Orthodox. Then some kind of congregational church, and then a synagogue. I believe it's Methodist at the present time. Though I'm not sure of that. . . . "

The man's incredible, thinks Nelson-Rees's prisoner, he even spells out what he doesn't know.

"These are all original little Victorian houses. Bret Harte lived and worked in this area. He had a place with his stepfather here on Fifth and Clay, his stepfather who was the model for the character of Colonel Starbottle in his short story "The Romance of Madrono Hollow." Do you know Bret Harte, the writer?"

The passenger hesitates, then nods, wondering whether a nod will bring momentary relief or only encourage his captor.

But it makes no difference. Nothing can stop Nelson-Rees now. He is free associating in high gear, interrupting himself at every turn, improvising around the one tune his brain is always humming: Origins. Histories. Where things came from and what they used to be.

It wasn't just buildings and roadways. He read postmarks, for instance, as if they were clues to buried treasure. If a letter arrived in his office without one, he would take out his red felt tip pen, circle the stamp boldly, and write "NO CANCELLATION!" across the envelope before filing it away. It really bothered him not to know where and when the thing had been mailed.

On napkin strips and pages torn from scientific journals, he often wrote up little summaries of conversations he'd had. Some of his notes of telephone calls even specified that he'd been called collect. These along with letters, telegrams, memos, conference agendas, lab notes, articles, and scrawled copies of questions he had asked speakers at certain meetings—as well as the speakers' replies—all of these were stuffed into thick, black notebooks that filled the bookshelves in his office. As one of Nelson-Rees's close friends once explained, "When Walter says something to anybody, he's got a piece of paper with his name on it, with the date, and what was said. It's there. You can criticize it. You can impugn it. You can say it's a lie. But it's there."

It was the same with his interest in art. Provenance is the term for a painting's history, and until he knew the provenance in full detail, Nelson-Rees felt he couldn't appreciate the work. When was it painted, where was the artist working at the time, who were his teachers, was it ever exhibited? It was probably under the force of this compulsion in 1953 that he had changed his very name, from Rees to Nelson-Rees, thereby specifying both parents, his own provenance. And so it was with cell cultures. If you didn't know everything about the cell line you were experimenting on, *especially* its true identity, then you didn't know anything.

So it was probably inevitable that Nelson-Rees would publish a second list of HeLa casualties. As it turned out, though, there were other good reasons.

Since the first article, the one that had "indicted" five cell lines as HeLa, he and Flandermeyer had discovered eleven more HeLa-contaminated cultures. When Nelson-Rees told the people who established these cultures that the cells were spoiled, many simply refused to believe him. It was not enough to deliver the bad news to these researchers and hope that, like Bob Bassin, they would do the honorable thing. As each of the eleven turned up, it became increasingly apparent to Nelson-Rees that he would have to warn the world himself through another published list.

SH3, the purported breast cancer culture established by Gabriel Seman at the M.D. Anderson Hospital, was one such case. Seman consulted a karyologist at Anderson who disagreed with Flandermeyer and Nelson-Rees's judgment that SH3 had HeLa chromosomes, the conclusion with which Miller had concurred. Seman also changed his story about the origin of the culture. After examining a photograph in his files, he concluded that the woman who donated the cells was not Caucasian as he had originally reported, but Mexican. Since the rare type A enzyme occasionally appears among interracial Mexican-American populations, according to Seman, this meant SH3 was not necessarily a HeLa culture.

Having satisfied himself that SH3 was a bona fide

breast culture, Seman saw no reason to curb its use or warn colleagues. While Nelson-Rees was preparing his new list, which would blow the whistle on SH3, the cell line was already being used by breast cancer researchers at the Memorial Sloan-Kettering Cancer Center in New York and by a group in the Soviet Union, as well as by an investigator who worked separately from Seman at Anderson. Indeed Seman would soon publish his own report in the journal *Cancer*, officially announcing the availability of this new breast cancer culture to the scientific community at large. In the article, he would rule out the possibility of HeLa contamination.

ElCo was a cell line that Nelson-Rees chose to include on his second list for similar reasons. According to oncologist Roland Pattillo, the man who established it, ElCo was yet another breast cancer culture. In his laboratories at the Medical College of Wisconsin, Pattillo had used the cells to test the effects of certain chemotherapy drugs before administering them to the woman from whom the cells had supposedly been taken. He also studied the reaction of other breast cancer patients to this "breast culture" using a skin test something like the ones used to check for allergies. These patients showed an immune reaction to ElCo as if their bodies had already been primed against it, suggesting to Pattillo that different breast tumors might have certain characteristics or antigens in common.

The only problem, Nelson-Rees told Pattillo after checking the culture, was that ElCo was neither a breast cancer culture nor representative of Pattillo's patient. It was a HeLa culture.

Pattillo protested that breast cancer patients would never have had an immune reaction to the cells unless they were genuine breast cancer cells. He would later concede, however, that he never tested that assumption by checking patients' reactions to HeLa cells. Pattillo also explained that he knew all about the infamous tumor of Henrietta Lacks, having been an associate in the laboratory of George and Margaret Gey soon after Mary Kubicek placed that fateful bit of tissue into the roller tube. But he said he had no HeLa cells in his own lab, so contamination was impossible.

Nothing in his lab was *labelled* HeLa, that was true. But he had been experimenting with another breast culture, Bassin's HBT3, and apparently HBT3 had dropped in unexpectedly on some of the other cell lines in the neighborhood. Nelson-Rees and Flandermeyer's analysis showed that ElCo contained the very same banded chromosomes as did HBT3 and its HeLa-contaminated cousins: the four standard HeLa markers and the newer pair, the Mickey Mouse and Zebra.

Pattillo just didn't believe it. He kept working with ElCo as if it were a breast tumor culture and kept publishing results. "That's what really got the wheels rolling around here," Flandermeyer recalled. "That's the kind of thing that got Walter all fired up and publishing new lists and writing letters to editors."

But what worried Nelson-Rees more than Pattillo's reaction was the fact that HeLa cells were still circulating in their old HBT3 disguise. HBT3 had been on the first list in 1974. For the last two years he had been making as much noise as he could about HBT3 and the four other HeLa-contaminated cultures he had originally stumbled upon. And yet here was Pattillo claiming he had no HeLa in his lab—although, yes, he did use a little HBT3 now and then.

Unfortunately, HBT3 was not the only active alumnus from the first list. Nearly everyone, it seemed, was still using MA160, the bogus prostate cell that by the mid-1970s had helped researchers waste an estimated $200,000.

MA160 was developed in 1966 through the collaboration of several scientists and a man named Monroe Vincent, a partner in the biomedical supply firm of Microbiological Associates, Inc. The scientists provided some of the technical expertise; Monty Vincent personally contributed the cells, a lump of tissue that had been taken from his own prostate. The lump was biopsied by his doctor who was concerned that it might be malignant. Fortunately, it turned out to be benign.

It also turned out to be a good source of culturable cells, or so it seemed. The cells from Vincent's prostate languished in the laboratory for a few months and then, touched by the miracle of "spontaneous transformation," were reborn as a bunch of frantic cancer cells. The first long-lived line of prostate cells, MA160 became a bestseller.

As it happened, though, the cells were exactly the opposite of what they were supposed to be. Not male, but female. Not prostate, but cervix. Not from a white donor, but a black one. MA160 was arguably the most tasteless of HeLa's practical jokes.

Stan Gartler was the first to nail MA160 as HeLa in 1968. He added it to his list of the original 18 HeLa-contaminated cultures based on its having the A type of G6PD enzyme. Vincent and his co-cultivators, however, dismissed the finding in their first official description of the cell line in *Science* in 1970. They theorized that Vincent might have had Negro ancestors from whom he inherited the rare, black enzyme. They added that the cells contained Y chromosomes, proof they had come from a male donor.

In 1973, however, Ward Peterson, Nelson-Rees's regular partner in Detroit, reported that he could find no Y chromosomes even in the earliest samples of MA160 available. Furthermore, having tested Vincent's own blood, he could say without a doubt that whatever exotic ancestry Vincent claimed, he had ended up with type B G6PD, the type expected for whites. "His" cell line was certainly not his anymore, but that of a black woman. In 1974, Nelson-Rees and Flandermeyer found that MA160 not only had the type A enzyme and was missing its Y chromosome, it also displayed banded marker chromosomes identical to HeLa's. Another group that had been working with the culture, a team at the Anderson Hospital, no less, soon came to the same conclusions.

It wasn't long before Mukta Webber, the Colorado cell biologist, reported that MA160 failed to produce prostatic acid phosphatase, a chemical normally manufactured by prostate cells. Webber also reported that despite its unprostatelike behavior, MA160 was still a popular research culture, describing and giving references to the six recently published "prostate" studies based on the cell line.

But that was not the end of MA160. Early in 1975 a West German scientist named Frederick Schroeder announced the establishment of a new prostate culture at a conference in Italy. Nelson-Rees, who happened to be attending, heard Schroeder describe the new cell line, EB33, as quite similar to the prostate culture MA160, which Schroeder also had been studying in his lab. In certain respects, EB33, like MA160, behaved strangely, according to Schroeder. Nelson-Rees began asking questions. A few months later, having analyzed several samples of EB33, he wrote Schroeder that his new prostate culture was definitely MA160, which was definitely HeLa.

And still MA160 and its alter ego endured. Many researchers went ahead and worked with EB33 just as others had continued to use MA160. Some, including Schroeder, used both cultures in their experiments, presumably to increase the validity of the observations. Here it was again, Nelson-Rees's nightmare come true. Researchers were not only using MA160 as if it were a bona fide prostate culture, they were exposing all the other cell lines in their labs to a HeLa culture in disguise. Which was why in addition to warning people away from the eleven newly spoiled cultures like EB33, ElCo, and

SH3, Nelson-Rees decided his second list would have to remind them of the contaminated cultures he had publicized earlier.

In fact he decided not only to repeat the lines he indicted in 1974, but also to quote every report he could lay his hands on that branded any cell culture HeLa. Among others, he cited Gartler's original findings, recent studies by Ward Peterson in Detroit, and a paper soon to be published by researchers at the American Type Culture Collection. In the manuscript he submitted to *Science* accompanying this second list, he explained how widespread "secondary contamination" had become and admonished his colleagues, "HeLa by many other names can spell trouble."

The list, published in January 1976, was an impressive and distressing inventory. More than forty different research cultures had been commandeered by HeLa. If each culture had cost science a quarter-million dollars, as had the few for which damage estimates could be made, this list represented $10 million in losses. The evidence against each culture, organized in a detailed chart, was overwhelming too. Not only were these cells identical to HeLa on the basis of their G6PD enzymes and marker chomosomes, but many had also been tested for the enzyme PGM and for a few more biochemical traits called HLA antigens. These all matched HeLa's characteristics.

And, oh yes, there was one other piece of information under each cell line in Nelson-Rees's second list: a researcher's name. Any scientist who had supplied him with a culture that turned out to be HeLa was immortalized in what came to be known as Nelson-Rees's hit list. No one considered this an honor. A few felt deeply betrayed and never spoke with him again. And the ranks of those who thought Nelson-Rees was just out to make a name for himself grew considerably.

Jim Duff and some of the other bureaucrats at the institute wanted to know why he had to name names. "You keep pushing people," Duff complained. Why, they wondered, couldn't Nelson-Rees simply give the sample's "passage level"—the number of times it had been grown out, cut up, and transferred to new flasks, which would give a relative indication of how close it was to the original culture—or identify it in some other way, and leave off the scientist's name?

Leave off the name? That, to Nelson-Rees, was like—well, like asking the Post Office to forget about the postmark. Why not take down all the road signs and let drivers fend for themselves? Why not ask the telephone company to publish just the phone numbers and have people guess to whom they belong? To leave off the names would defeat the whole point of his list. He was trying to be as precise and complete as possible, to specify where the HeLa-contaminated cultures had come from and, by implication, where they had *not* come from. After all, he hadn't tested every sample of these cell lines. Not yet.

"After the 1974 paper," Nelson-Rees explained to *Science News* magazine, "some researchers analyzed cultures they had been using of the same type we 'fingered' and found them to be bona fide bladder carcinoma cells, or whatever. Therefore, the source of these cultures becomes an important piece of data. I felt obligated to state from whence these cultures came and let the other shoe drop where it may."

Of course, Nelson-Rees must have known that naming names was, as the magazine put it, "an action sure to be interpreted by some as unfriendly." But maybe he figured that in order to make any progress, the second list had to be more combative than the first.

"At this point," he told *Science News*, "I'm going to go hide."

Naturally, he did nothing of the kind.

11

Another Run-in With Relda

Nelson-Rees was browsing through the April 1977 issue of the *Journal of the National Cancer Institute* when he caught the scent of something mildly suspicious on page 863.

Now some people would have said he went looking for bad news, especially after that hit list with the names on it. But the way he saw it, he just wanted to make sure the reports of good news were genuine. And starting on page 863 of the institute's journal there appeared to be very good news indeed, a report of some promising findings about lung cancer.

Richard Akeson of UCLA medical school had found several specific antigens associated with the cells in a lung tumor culture. Antigens are like biochemical dog tags, tiny nameplates on a cell's surface that can be recognized by the body's immune system, though in the case of cancer the system seems unable to successfully attack even when it recognizes the enemy. While the discovery was preliminary, Akeson wrote, such tags might eventually help doctors diagnose this particular kind of lung cancer more effectively, perhaps earlier, and help them monitor its progress. Furthermore, though Akeson didn't actually mention it, there was always the hope that if tumor-specific antigens could be found for certain cancers, they might serve as signposts to be followed by cancer-killing drugs or other selective assaults to be developed in the future. Such was the promise of an antigen found to be unique to a particular kind of tumor cell.

Nelson-Rees knew from experience, however, that antigens specific to a certain type of cell—and only that type—were practically impossible to find. And, of course, finding a unique antigen wasn't enough. You also had to be certain of the tumor cell's identity in the first place, so that later when you saw that antigen on the cells of an unidentified culture you could say, for example, aha, we must have a dishful of Russian bladder cancer here. Given the proclivities of biologists to mix things up, Nelson-Rees figured, any tissue-specific claim was worth a closer look.

According to the "Materials and Methods" section of Akeson's report, the lung tumor cell line was called 2563 and had come from Litton Bionetics, Inc., in Kensington, Maryland. Following that bit of information was a footnote that directed Nelson-Rees to the bottom of the page. There amid the fine print, quite unexpectedly, he bumped into none other than Relda Cailleau. It seemed that 2563 was another name for MAC-21, the lung culture that had triggered the hotel lobby screaming match in Montreal two years earlier. Despite Cailleau's assurances at the time that MAC-21 was not HeLa, the matter was left very much unresolved in Nelson-Rees's mind. Now that someone was basing important new claims about lung cancer on this twenty-year-old culture, he decided the question had to be settled.

He ordered samples of 2563 from Litton Bionetics that summer, and by September the verdict was in. According to Ward Peterson's analysis, the culture's G6PD enzyme was type A, the same as HeLa's. Although Cailleau had taken the original MAC-21 cells from a fifty-three-year-old man, Nelson-Rees and Flandermeyer saw no male Y chromosomes. And in nearly every cell, they found abnormal chromosomes whose shapes and banding patterns matched three of HeLa's best-known markers. Nelson-Rees wrote a detailed letter to John C. Bailar III, editor of the *Journal of the National Cancer Institute,* asserting that 2563 had been contaminated by HeLa and was not at all characteristic of lung tumor cells. He asked that the letter be published as soon as possible, saying his results were "an important addition to the increasing knowledge of what does and does not constitute organ-or tissue-specific antigens."

Three months passed with no word. Nelson-Rees called the journal, and someone there explained they had sent copies of his letter to Akeson and Cailleau so that they could write rebuttals to be published along with his letter. Akeson's reply had come in, a cautious concession that the "lung cancer cells" were apparently HeLa, but they were still waiting for Cailleau's.

Convinced as ever that the cells were pure, Cailleau had pulled out the one remaining ampule of MAC-21 in her possession and had it analyzed by Sen Pathak, a chromosome banding expert at the M.D. Anderson Hospital, and Michael Siciliano, the local enzyme man. She was "disappointed," as she later described it, when Pathak reported seeing HeLa's marker chromosomes in MAC-21 and Siciliano said not only was the G6PD HeLa's type, but thirteen other enzymes he tested matched as well.

Disappointed, but undaunted, Cailleau still insisted that MAC-21 had been a genuine lung cancer cell for many years, and if it was HeLa now, well, that was somebody else's doing. She explained to her colleagues, as she would afterwards to anyone who asked about MAC-21, that she had shared the cells with other researchers in San Francisco before coming to Houston. When she moved she took with her two different batches: one of live samples from those researchers and another of frozen samples that only she had used. Then in the early 1970s some fool technician had failed to keep the freezer stocked with dry ice, and she lost seventeen years' worth of cell cultures, including the bona fide MAC-21 cells. Her only remaining samples were those that she had had no direct quality control over.

"The problem," Cailleau said, "is that often contamination of cells is a secondary or a tertiary thing done by associates or technicians, and the originator knows damn well he or she didn't contaminate them. In my hands, there never has been a HeLa contamination." In other words, yes, the MAC-21 cells she had now were indeed HeLa, but it wasn't her fault. The same was true for the samples she had sent a few years earlier to a variety of colleagues, including one who must have passed them on to Litton Bionetics under the name 2563, some of which had ended up in Richard Akeson's lab.

All this Cailleau explained in a letter to Bailar at the *Journal of the National Cancer Institute* in January 1978. Painful as it must have been, she also telephoned Nelson-Rees—collect—and laid out the whole mess for him. She sounded uncharacteristically subdued, Nelson-Rees thought, even a bit sad. As pleased as he was to hear his suspicions confirmed, he kept his I-told-you-so's to himself.

Still, it would be another six months before anyone but Nelson-Rees, Akeson, Cailleau, and Bailar had any inkling that MAC-21 and 2563 were not what they were advertised to be. Scientists who had read Akeson's report on lung tumor-specific antigens would have had an entire year to follow his lead without any indication it might be a dead end. Finally Nelson-Rees's letter appeared in the journal's June 1978 issue. It was followed by a letter from Akeson who said that based on Nelson-Rees's work "it seems reasonable to presume" the cells he had thought were from a lung tumor were actually HeLa cells. He added, however, that some of the antigens in these cells *did* seem to resemble those seen in genuine lung tissue, concluding that further studies were needed to sort through the confusion. Following Akeson's letter was a one-paragraph statement from Cailleau conceding that the single culture of MAC-21 she had checked was HeLa. "I am certain that the original MAC-21 was obtained from a mucoid adenocarcinoma of the lung and remained uncontaminated for several years," she wrote with a trace of controlled testiness. "The addition of HeLa cells to our cultures occurred later and the transfers were carried out by several people."

Around the time of this embarrassingly public cell mix-up, Nelson-Rees received a draft report from a microbiologist at the National Cancer Institute. It was an extensive review of breast cancer cultures available to experimenters, and, as Nelson-Rees was so experienced in telling good cells from bad, the microbiologist asked him to critique the report before it was submitted for publication. Oddly enough, there were several cultures listed whose technical descriptions were missing. The cover letter explained these were Cailleau's cultures, and Cailleau had asked that all information on them be deleted from any copies of the draft that Nelson-Rees might see.

Nelson-Rees was amazed—first that Cailleau would have the gall to dictate the handling of someone else's manuscript and, second, that the author would knuckle under. When Nelson-Rees called to find out more, the microbiologist said that Cailleau had been "adamant on this point; abusive, in fact."

Obviously Cailleau had been more than just miffed by her latest dealings with Nelson-Rees. She was determined to make the MAC-21 incident the last time he would ever stick his nose into one of her cell cultures.

A few weeks later, the Tissue Culture Association asked Nelson-Rees to evaluate a stack of abstracts, or summaries, of research reports submitted for the association's upcoming annual meeting. He found one of the abstracts particularly interesting. It described an unusual sample of breast tumor cells that exhibited a surprising number of characteristics similar to HeLa's. In fact, many of its chromosomes looked like HeLa's banded markers, the authors said. Yet they knew for certain the cells were not HeLa because they had come directly from the patient—out of a pleural effusion, a sample of liquid taken from the patient's chest cavity, to which the tumor cells had spread. The cells in the effusion were analyzed immediately; they had not sat around the lab and had a chance to become contaminated. The implication was that in this case the standard techniques of identifying a HeLa cell might well have led to a false diagnosis. These bona fide breast cancer cells might have been wrongly branded as having been overtaken by HeLa.

The abstract had been submitted by three scientists at the Anderson Hospital: Sen Pathak, Michael Siciliano, and Relda Cailleau.

Nelson-Rees almost laughed. He doubted the claim, of course, but was intrigued at the same time. In part because he wanted a chance to examine the supporting data, he recommended that the report be accepted and presented at the meeting. He then wrote to Pathak, the karyologist, asking to see photographs of these curious chromosomes.

When the photos arrived two months later, eagle-eye Flandermeyer looked and looked but he couldn't see any known HeLa markers among the banded chromosomes. Nor could Nelson-Rees. They sent copies to K.S. Lavappa, the chromosome expert at the American Type Culture Collection, who wrote back that there certainly were some strange-looking chromosomes in these cells, but none of them were HeLa markers.

Two weeks before the annual meeting, at which Pathak was scheduled to discuss this cell line, Nelson-Rees received yet another manuscript, this one from Bailar, who asked him to review it for the *Journal of the National Cancer Institute*. It was a much more detailed report on the very same cell line. In addition to Pathak, Siciliano, and Cailleau, T. C. Hsu, head of the Tissue Culture Department at Anderson and described by some as the guru of cytogenetics, had put his name on the man-uscript, adding considerable credibility to it.

What the earlier abstract had only implied, this twenty-five-page paper announced loud and clear: that the standard tools of HeLa hunting—enzyme tests and chromosomal analysis—were of questionable value. What's more, it suggested that every HeLa case Nelson-Rees had ever published was now in doubt. In short, it looked like the gang from Anderson was indicting the in-dicter himself.

"The HeLa markers have been extensively used by investigators to identify . . . cell line contamination," the authors wrote, citing references to three of Nelson-Rees's published reports, including the latest hit list. Because they found chromosomes similar to HeLa markers in this tissue sample that was clearly not HeLa, they concluded that "marker chromosomes alone are not unequivocal evidence for identifying cellular contamination."

Wait a minute, thought Nelson-Rees. From what he and Flandermeyer and Lavappa could see, *there were no HeLa markers in these cells.* Sure, maybe there were a few that looked vaguely similar. Maybe someone with an untrained eye might be confused. But there was not a single marker chromosome in the photographs that a qualified karyologist should mistake for HeLa's. And even if there were, Walter Nelson-Rees would never label a cell culture HeLa based solely on marker chromo-somes. There were also the enzymes to consider.

But the folks from Anderson observed that twelve out of fourteen isoenzymes tested in their breast cancer cell were identical to HeLa. "The suggestion, then, that cell lines showing only three enzyme characteristics in common with HeLa be considered *de facto* strains of HeLa may be overly simplistic," they wrote, offering footnotes to two more of Nelson-Rees's reports.

Overly simplistic? Nelson-Rees couldn't believe this stuff. He'd never claimed that enzymes alone are enough. But *if* the marker chromosomes match *and* you pick the proper enzymes to check, then three can be plenty. The single finding of type A G6PD, for instance, in a culture thought to be from a white donor is more than enough to tell you something is awry.

Ah, but the "HeLalike" cells reported in this manuscript had come from a black woman who carried the A form of G6PD. How convenient, thought Nelson-Rees.

As he scanned the rest of the report, he noticed the authors neglected to point out that many of the other enzyme variants they tested were common to 90 percent of the human population, including the late Henrietta Lacks, and therefore of relatively little value in distinguishing their breast cancer cell from HeLa. And by a strange coincidence the two enzymes that turned out different from HeLa's happened to be the very last two they tested, even though one of them, called PGM3, had become a standard test for HeLa, and therefore likely to be among the first checked by anyone really interested in finding out if he had a HeLa cell or not.

The handling of the enzyme data was not the only thing Nelson-Rees found suspicious. He called Flandermeyer in and they pored over the chromosomal mug shots Pathak, Cailleau, and colleagues had used to prove their case. In an hour or so they were both shaking their heads, astonished.

The mug shots were photographic composites, constructed from bits and pieces of various photos of various cells in this breast cancer culture.

When you're comparing banded chromosomes from two different tissue samples—a known HeLa culture, say, and one you suspect may be contaminated with HeLa—there's only one correct way to do it, according to Nelson-Rees and most other karyologists. After searching hundreds of photos you begin to recognize the same complex of markers, the same group of defective chromosomes, appearing in every one of your suspect cells. You then pick a single cell in which this group displays its banding patterns clearly, cut out the defective chromosomes, and put them next to markers from a HeLa cell you've analyzed and photographed. In a given side-by-side comparison, the mug shots must come from only two individual cells. If you have to pick one chromosome from this suspect cell and another from that in order to find ones that match those from the HeLa cell, then you're reaching.

Apparently that's just what the Anderson workers had done. After a close look at all the photographs, Nelson-Rees and Flandermeyer were convinced that the comparison photo didn't display mug shots of chromosomes from only two cells, one from the breast culture and one from a HeLa line, but from at least seven, four different breast cells and three HeLas. What's more, the HeLa mug shots appeared to have been taken from another researcher's published report, not a culture the Anderson workers had studied themselves. In fact, they had cut out and reassembled so many mug shots in order to make their comparison look convincing they had even pasted in the image for one of the breast chromosomes upside down and reversed.

So this was what Pathak was going to unveil at the annual meeting. And as if that wasn't enough, the Anderson gang wanted to publish the entire argument in the *Journal of the National Cancer Institute* as well. Nelson-Rees was astonished that anyone would go to such lengths to discredit him.

For the next few days he drove Flandermeyer and the others hard. He demanded a detailed analysis of the mug shots showing that they weren't HeLa markers despite the photographic sleight-of-hand and offering alternative descriptions for each chromosome. When that was done, Nelson-Rees sat down, poison pen in hand, to draft a statement he planned to deliver at the annual meeting following Pathak's presentation. When it was done, he took it to Flandermeyer, who recommended toning it down. They removed a few of the more offensive passages, including a final statement Flandermeyer felt was particularly undiplomatic. Nelson-Rees then returned to his office and wrote the venom back in.

12
Showdown

As soon as Nelson-Rees arrived at the Denver Hilton Hotel, where members of the Tissue Culture Association were gathering for their twenty-ninth annual meeting, a friend came up and said, "The Anderson group is out to get you."

The following morning, the inflammable Relda Cailleau ignited at the mere mention of his name. She was at a meeting of the association's executive council, where she learned that Nelson-Rees had been elected vice-president of the association and was to be officially installed the next day—and she couldn't contain the fire and fury.

"My god," she blurted out. "How could he have been elected vice-president?"

She told her startled fellow council members how Nelson-Rees had attacked her work and that of many well-respected scientists, how his methods were antagonistic and his motives destructive. "His approach is always, 'I've got to prove I'm right and you're wrong,'" she railed. *"My God, the man is paranoid!"*

Cailleau urged the council to demand his resignation, a difficult task as Nelson-Rees didn't yet hold the office. But she had made her point, several times over, and set the stage for the main event later in the day.

Nelson-Rees was set himself. Earlier he had told Bob Stevenson, "I'm really going to throw the fat on the fire."

"Think about it," Stevenson answered.

Some 200 people packed into the Hilton's Vail Room that afternoon, considerably more than were originally expected for the session on mammary tissue. They had heard there was quite a show in store. Some said it might really get ugly. They came not necessarily to root for one side or the other, but—like decent townsfolk gathering on rooftops and peering out of windows at a gunfight in the street—just to watch.

The paper by Pathak, Cailleau, and Siciliano, entitled "Fresh Pleural Effusion from a Patient with Breast Cancer Showing Characteristic HeLa markers" was scheduled third. The audience sat restlessly through the first and second talks. Then, about thirty minutes into the program, the chairman read the title of the paper, and Pathak, the young Indian-born chromosome expert, came forward to deliver the talk. Cailleau was sitting in the audience about halfway to the back of the room. Nelson-Rees was about ten rows behind her. Siciliano had stayed home in Houston.

Pathak seemed apprehensive. He spoke a bit shakily in that British-educated, Indian accent, painfully formal and polite. Whenever he mentioned a colleague, it was always by title and full name, pronounced in one breath as if it were all a single word: "We examined chromosomes in cells obtained by DoctorReldaCailleau. . . . Isoenzymes from DoctorReldaCailleau's cells were analyzed by DoctorMichaelSiciliano."

Pathak described the findings in a straightforward way: These cells taken directly from a breast cancer patient showed chromosomes that were quite similar to HeLa's characteristic marker chromosomes, he said. And thirteen of fifteen isoenzymes checked were identical to HeLa's. (They had tested one additional enzyme since writing the draft report Nelson-Rees had seen.) There were no direct references to Nelson-Rees, no outright challenges to the usefulness of his techniques as there were in the written version of this talk that had been sent to the *Journal of the National Cancer Institute*. Still, to some members of the audience, especially to Nelson-Rees, the insinuation was clear.

When Pathak had finished, Nelson-Rees raised his hand. The session chairman, having heard that trouble was afoot, quickly called on someone else. Nelson-Rees decided he would not be denied again. When that question was answered, he simply stood up and without the aid of a microphone began to read his statement.

"As a referee of this abstract for the program committee, I questioned the use of the adjective 'characteristic' in the title, followed by 'similar' in the text. Dr. Pathak subsequently sent me a photograph of the 'marker' chromosomes stating he 'was puzzled regarding this observation.' I am puzzled as well, not with the observations but with the way they are being presented."

The chairman tried to cut him off, explaining that the presentation had gone over its allotted time.

"John," said Nelson-Rees as he moved from his seat up towards the front of the room, "I'm going to finish this."

He went on to explain that as a reviewer of the manuscript submitted to the cancer journal, he had looked over all the photographic evidence the authors used to prove their case. He said he discovered they had "fished about" in several different cells from this breast tumor to find their so-called HeLa chromosomes, a highly irregular procedure.

"*We* have carefully studied many different cell lines and observed a good number of marker chromosomes, but we've *never* had to create a collage such as has been done here in order to sensationalize a point."

The audience was aghast at such a flagrant breach of etiquette. One simply didn't use such language. Perhaps a point would need to be confirmed; maybe there was a chance for an alternative interpretation. But publicly accusing learned colleagues of fishing about for data and sensationalizing their results—Nelson-Rees might just as well have been spoon-launching mashed potatoes at a formal dinner.

"As a matter of fact," Nelson-Rees proclaimed, "without stating so and without giving credit to the published work, the authors lifted HeLa marker chromosomes from the work of Heneen which best matched their altered chromosomes and presented them in the manuscript apparently as their own observations." He held up the Swedish journal *Hereditas* that contained the report he claimed they had "lifted" from and added that despite such pilferage, the chromosomes they presented are not convincing matches.

There was a sudden disturbance in the audience. Relda Cailleau was becoming agitated, muttering loudly to the people around her, but Nelson-Rees pressed on.

"Aside from the above-mentioned irregularity in scientific procedure, it appears that some normal chromosomes were erroneously classified as markers and vice versa, and at least in one instance a chromosome's image was used twice, albeit after reversal of the negative."

Nelson-Rees explained that he had sent photographs of the "so-called characteristic" HeLa markers to Lavappa for a second opinion and quoted Lavappa's written reply: " 'There is no trace of the four distinct HeLa markers, which are well characterized and have appeared in many publications.' " And then it was time for the finale, the part that he had typed out in capital letters.

"I suggest you stick to pleural effusions," he said to a dumbstruck Pathak at the podium, "and leave the HeLas to us."

A momentary silence.

Pathak broke the trance and added a touch of absurdity to the scene by thanking Nelson-Rees profusely. Nelson-Rees dropped 150 copies of his statement in a pile at the back of the room and left. Suddenly pandemonium broke out over the Vail Room. There were five more speakers scheduled for the afternoon, but half the audience was rushing for the exits.

Relda Cailleau made her way over to the stack of statements, grabbed about thirty of them, and headed out. In the hallway she worked the crowd, offering copies of Nelson-Rees's attack as proof the man had no business reviewing manuscripts for technical journals (Just look at how he violated the reviewers' oath of confidentiality!) or serving as vice-president of the Tissue Culture Association (Was such rude, unethical behavior fitting for an officer of our group?).

Pathak returned to his hotel room, where he remained for most of the last two days of the conference. He would later explain his absence by saying he was deeply shaken by the wrongful accusation and public humiliation he had suffered. In fact, he was so distraught that he refused to do any research on the chromosomes of tumor cells for an entire year. Long after the incident, Nelson-Rees offered an alternative explanation: "A little dog—or even a big one—who has peed on the living room rug and is found out *always* hides underneath the couch for a couple of hours."

At the moment, however, Nelson-Rees was fresh out of quips. His feistiness was suddenly spent. He was exhausted and, he noticed, quite alone. As he moved through meetings the rest of the day, people avoided him, looked away disapprovingly when he approached, and whispered about what a disgraceful performance he had given.

One of those who attended his performance and had been especially shocked was Colonel Albert Leibovitz. A tissue culturist formerly with the U.S. Army and now at the Scott and White Clinic in Temple, Texas, Leibovitz had at one time respected Nelson-Rees's crusade to clean up the field, or so he said. Recently, though, it looked to Leibovitz like Nelson-Rees was on a maniacal rampage.

His means were extreme. And what he had done this afternoon could only be described as scandalous. "It is *so* unethical to be a reviewer of a paper and use confidential information to tear its authors down," Leibovitz declared.

Leibovitz, of course, may have felt some special sympathy for the Anderson crew, having had a recent tussle with Nelson-Rees himself. The matter involved eight cell lines Leibovitz had established: colon, bladder, and six other kinds of cancer, all of which had apparently been taken over by a single line of colon cancer cells. Although it was not a case of HeLa contamination it was still a large mix-up, and Nelson-Rees felt it ought to be publicized immediately. When Leibovitz balked, preferring to double-check the findings and investigate the source of the contamination further, Nelson-Rees tried to publish a letter in the *Journal of the National Cancer Institute,* warning other scientists away from these cells. Leibovitz was furious. Certainly Nelson-Rees had been right to want to get the word out, but he was jumping the gun for no reason, Leibovitz believed, and worse, he had broken an implicit confidence—just as he had in attacking the report by the Anderson group.

Later that afternoon at the general membership meeting of the Tissue Culture Association, it was Colonel Leibovitz who stood up and with great indignation urged that the association create an ethics committee to deal with one of its members who, at a presentation earlier in the day, "committed unscientific acts without presenting proof to support his statements." This person's verbal assault was all the more inappropriate, the colonel explained, because he had been the confidential reviewer of a manuscript covering this particular presentation.

Seated out in the audience, Nelson-Rees kept a faint, nervous smile on his face as the members discussed the need to discipline the unnamed outlaw and eventually voted to have Leibovitz draft a resolution for consideration at the next meeting of the executive board.

That night Bob Stevenson took Nelson-Rees out for a drink and a little lecture. It was another version of the friendly advice Stevenson had regularly offered: Diplomacy gets you farther. When you do something, you do it properly, or you risk losing credibility.

On the other hand, Stevenson added, he was only too glad somebody was stirring things up. He grinned his mischievous grin and said, "It's always nice to drop a turd in the punchbowl once in a while."

Then came the letters. In a few weeks Keith Porter, the president of the association, was hip deep in testimonials of outrage, shock, and dismay:

> I am writing to express my concern about the incident that took place in Denver. . . . Nelson-Rees has abused his privilege as a reviewer . . . and impugned the integrity of Drs. Pathak et al. I hope the Executive Board will treat this matter with the utmost seriousness, because I believe the credibility of the Board, as well as the good reputations of Drs. Pathak, Siciliano, and Cailleau are at stake.

> I am but one of many who was shocked and offended by the serious and unethical activities of Walter Nelson-Rees.

> Dr. Nelson-Rees behaved ungentlemanly and indiscreetly. He appeared to lack the objectivity and open-mindedness of character important for a good scientist.

Leibovitz's resolution also arrived, four pages documenting Nelson-Rees's crimes. "It is immaterial whether his findings are right or wrong," the statement read and urged that he be censured, stripped of his vice-presidency, and banned from reviewing manuscripts for the association.

It was all very well, all this righteous wrath, but as far as Porter could see, there was no way to satisfy it. Nelson-Rees was guilty of no malfeasance, and even if he were, he was not in office when he so distressed everyone; that was the day before he was officially installed.

When the officers, including Nelson-Rees, met a few months later, Nelson-Rees agreed to apologize to Pathak in writing, but only for the manner in which he "disagreed" with Pathak's report. The president hoped that would satisfy the offended.

It didn't come close to satisfying Relda Cailleau, who wrote the president four months after the incident:

> Dear Dr. Porter,
>   If he [Nelson-Rees] and other members of the Council consider that his apology for having "offended" Dr. Pathak is enough to close the subject, they are mistaken. His *unethical* conduct is not mentioned and it is the only subject of importance for the T.C.A. and for any reputable scientist. . . . If he cannot be removed from office or forced to resign because the T.C.A. constitution has no provision for such a contingency, he must at least be made to acknowledge his unethical conduct (which incidently is *not* limited to this occurrence). . . . He should no longer be allowed to review articles without some safeguard to the people whose reputations he tries to destroy. The only way this can be done is to inform the membership as well as the editors of the various journals in which his comments have appeared.

Porter answered:

> Dear Relda:
>   . . . Unofficially, I would urge you to forget the incident. I have heard some very abusive attacks of one scientist on another at scientific meetings and they did not become the business of the society or institution sponsoring the meeting to settle. The TCA cannot undertake to censure, punish or even reprimand its members for "unethical" behavior. As I have said before, Walter was duly elected and cannot be removed from office without going back to the membership of the Association. I do not think that you could influence the required plurality.

What amazed Nelson-Rees was that nowhere amid the outcry over his Denver recital was there any discussion of the merits of his argument, no effort made to determine whether the results reported by Pathak and company were trumped up, as Nelson-Rees had charged. They had claimed that his methods for fingering a cell line as HeLa were in doubt and, by extension, his hit lists and other reports were wrong. Well, Nelson-Rees couldn't stand being told he was wrong. And it incensed him that no one cared to consider the substance of his rebuttal.

After the Denver meeting, he sent off a stinging critique to Bailar at the *Journal of the National Cancer Institute*, restating and amplifying his argument. The Pathak report, he wrote, should not be published unless its "entire concept is altered." He questioned its "sincerity and merit" and accused the authors of resorting to "photographic trickery" in their presentation of the chromosome evidence. As for the enzyme data, he wrote, it may well be true that thirteen out of fifteen enzymes tested were the same as HeLa's; but two of them, the two most useful for identifying HeLa cells, *were different*. Hence the system holds up. No one would mistake these cells for HeLa, as Pathak and Cailleau's gang had claimed.

Like his speech in Denver, however, these criticisms were ignored. The paper, barely revised from the original version, was published in February 1979.

When they were asked about it years later, the Anderson researchers insisted they had never set out to discredit Nelson-Rees's years of crusading, as he had somehow imagined. It had all begun as a simple study of chromosomes in fresh tissue samples. Pathak wanted to find out if the defects observed in chromosomes of cultured cells resulted somehow from extended growth in a laboratory dish. Or did cancer cells taken fresh from a patient show abnormal chromosomes as well? Cailleau had been supplying him with fresh pleural effusions. And one day they simply stumbled across this unusual bunch of cells containing chromosomes similar to HeLa markers.

T.C. Hsu, the senior cytogeneticist in the lab, thought the finding raised an important point in the current climate of HeLaphobia. In most cases, said Hsu, perhaps 98 percent of them, a cell that has HeLa markers is probably HeLa. But these markers may not be unique to HeLa; they may be created by some general mishap of cancer. So there may be a few cases in which a cell carries HeLalike chromosomes and yet is something quite different.

"The findings so shocked Nelson-Rees, he thought it rocked the foundations of all his work," explained Hsu. "But it didn't really rock his foundations."

"We were simply urging caution when diagnosing cells as HeLa," added Pathak, "not disagreeing with all his results."

Still, in their published report, and in follow-up articles and letters to certain colleagues, Hsu, Pathak, and Cailleau repeatedly warned against relying on chromosomes alone, referring to Nelson-Rees by name and suggesting that he had foolishly done just that—when he never had.

As for their need to create a collage, breaking with the traditional means of comparing chromosomes in cell lines, they explained that fresh tissue samples are always a mess, carrying all kinds of bodily by-products that are eventually filtered out of a long-term cell culture. So in a fresh pleural effusion, it is difficult to take clear photographs of all the marker chromosomes within a single cell.

"The composite was prepared in order to show an example that was clear-cut to the untrained eye," said Cailleau.

As Flandermeyer often pointed out, though, the identification of chromosomes entails a great deal of judgment. "There's an art to it, and if you're not careful you can incorporate your bias," he once said, thinking of the Denver affair. "You can make things conform to what you expect, especially if you pick around, choosing one chromosome from one cell and another from the next."

The presence of specific enzymes inside a cell is much less ambiguous, and therefore far more resistant, to an investigator's bias. On the other hand, the choice of which enzymes to look for and the order in which they are checked can create certain impressions.

Stephen O'Brien, an enzymologist at the National Cancer Institute and an ally of Nelson-Rees, was puzzled by the Anderson group's need to check fifteen enzymes in their breast cells before finding two that differed from HeLa's. Had he been doing the experiment, O'Brien said after reading the report, he would have analyzed perhaps six, and the two mismatches would have been among the first.

But according to Siciliano, the enzyme expert who collaborated with Hsu, Pathak, and Cailleau, the fifteen enzymes they checked were the standards, the ones they checked routinely. It just so happened the two that turned out to be different from HeLa's were tested last.

Like his colleagues, Siciliano insisted the report was not intended to discredit Nelson-Rees or to knock the foundation out from under his published findings. "I don't think there was any stuff in his published papers that's refutable," he said.

Why, then, the comment that Nelson-Rees's use of only three enzymes is overly simplistic?

"Most of the cells on his lists came from white donors," explained Siciliano, "so testing for only one enzyme, G6PD, was enough. But I was concerned for the future as more blacks donate tissue cultures, as they're doing in institutions like ours."

Motives, alas, are slippery fish. It is difficult for anyone to know what the Denver blowup was really all about. Anyone but Nelson-Rees, that is, to whom the truth was obvious.

"This group of people was trying to destroy years of

careful work and promotion of the use of good cell lines," he said after the incident. "It was just a case of this being put together to cause embarrassment to me.

"You know, I didn't 'impugn their reputations,' as some said of my rebuttal at Denver. I called them absolute shits. *They* were the ones who were unscientific and unethical, not me."

13

Even the Best of Labs

One of the standard comebacks to Nelson-Rees's appeals was, "Ah, but HeLa threatens only the second-string team. Only the minor league players are sloppy enough to mix up their cells. For those of us who know what we're doing, this is all quite irrelevant."

In October 1978, Jonas Salk, developer of the Salk polio vaccine, founding director of the Salk Institute—where some researchers practice philosophy as well as science—medical giant Jonas Salk stepped forward as one large exception. He let it be known that HeLa cells had taken over one of *his* cell lines.

Salk had actually learned of this HeLa takeover years earlier, but hadn't spoken of it publicly until that October during a meeting of several hundred cell biologists and vaccine experts at the Lake Placid Lodge in New York. The eminent Dr. Salk was describing a series of experiments he had conducted on terminal cancer patients in the late 1950s. The idea was to try to switch the patients' immune systems into combat mode by challenging them with an injection of foreign cells. The foreign cells came from a line of monkey heart tissue, the same line that supplied the cultures in which he had grown polio viruses to develop his famous vaccine. Salk believed these monkey cells had certain antigens, certain biochemical identification tags, in common with cancer cells. He hoped that by activating the patients' immune systems against the monkey cells, he might trigger an attack against the cancer as well.

Strangely enough, Salk told his attentive audience, where he injected the cells, some of his patients developed tumors. The distinguished scientist-statesman flashed a few slides of human arms carrying pea- and almond-sized abcesses, and he said calmly that in retrospect he believed those cells may not have been his harmless monkey heart cells at all, but HeLa cells.

Well, a nervous murmur filled the auditorium and there were a couple of gasps as Salk's colleagues tried to envision him making the rounds with Henrietta Lacks hiding in his hypodermic, a determined Jonas Salk shooting up dying patients with an extra dose of overactive cancer cells.

Salk added quickly that the ugly little knots dissipated in no time, most of them in three weeks, never to return. Apparently he hadn't helped most of the patients with these injections—although one man did undergo a remarkable remission and live another eighteen years—but neither had he given them any more cancer than they already had. And that was the whole point of his talk: to report that even when injected directly into human beings, HeLa cells do not cause cancer.

A rather bizarre conclusion to present, perhaps, except that this was a conference about the making of vaccines. In particular it was about the use of cells in making vaccines. For the last few decades, there had been a running argument in this crowd over whether so-called continuous cell lines—that is, established cultures that continue to reproduce themselves apparently without end—are safe "substrates," safe soil, so to speak, in which to grow viruses to be used in vaccines. Or was there a chance the mysterious forces that give these cultures their immortality might also trigger runaway growth—cancer, in effect—in the recipients of vaccine prepared in those cells?

People like Salk argued there were adequate methods of separating a vaccine's active ingredients from the cells they were prepared in, so that it didn't matter what kind of cells were used. In fact Salk believed that even if vaccines weren't filtered at all, even if whole cancer cells were injected directly into a human test subject, they would be rejected by the body and do no harm. Of course no researcher would try that experiment. Not intentionally anyway. But twenty years earlier he had done it accidentally, Salk told the meeting, and only later did he realize it. He had injected HeLa cells into a few dozen patients, and it hadn't bothered them a bit.

Some of those who heard this remarkable talk later played it down, saying it came as no surprise. Virologists especially and people who were in on the early vaccine work said they had heard reports long before October 1978 that Salk's monkey heart cell showed some HeLa-like characteristics. Nevertheless, that Salk had injected cervical tumor cells straight into human beings was disturbing to at least some of the conferees. There were also a few who thought Salk was awfully glib to conclude there was no danger in this simply because the lumps on patients' arms disappeared. Who knew what these HeLa cells had been up to inside each subject's body? The conference organizers apparently found this part of Salk's talk so unsettling they advised him to skip it in the written version to be submitted for publication. Sure enough, when the collection of reports was published the following year, Salk's paper made no mention of the inadvertent human experiments.

Nelson-Rees was one member of the audience who was openly shocked. To him, Jonas Salk had always been a renaissance scientist: brilliant, sophisticated, driven. That this "fantastic creature," as Nelson-Rees once described him, could be tripped up by HeLa tarnished the image considerably. On the other hand, here was a perfect demonstration of what Nelson-Rees had been preaching all along. Without eternal vigilance, HeLa can drop in on even the best of labs and the biggest of names.

Nelson-Rees, of course, had to be sure. Salk had said only that other scientists had reported his "monkey cells" to be HeLa; Salk had never checked them himself. So when it came time for questions, Nelson-Rees stood up and, as a few in the audience groaned and shook their heads, offered his services. If Salk still had samples of the twenty-year-old cultures, Nelson-Rees said, they could be clearly identified.

Salk graciously accepted Nelson-Rees's very public offer.

The following week Nelson-Rees dropped in on Salk at his institute in La Jolla, a sprawling six-story concrete labyrinth built around a courtyard, all of it set upon high cliffs overlooking the Pacific. He was shown into Salk's office, a wood-paneled sanctum on the fifth floor. Gulls and hang-gliders rode the clouds just outside the window. Nelson-Rees chatted briefly with Salk, picked up several samples of the cell line CH, cynomologous heart, and left for Oakland.

So, it does happen even to the first string, thought Nelson-Rees two months later, even to the titans of science. Salk's monkey cells showed no signs of monkey chromosomes. But, as Nelson-Rees said in a letter to Salk, they did contain five of HeLa's "well-publicized marker chromosomes" as well as the type A G6PD enzyme.

Salk wrote back thanking Nelson-Rees for his work. He added that he looked forward to follow-up discussions.

But the follow-up discussions never took place. Salk's reaction, in fact, wasn't very different from many of HeLa's less illustrious victims. Not that he fought Nelson-Rees's conclusion in the scientific press or even argued with him personally. But deep down, he just couldn't buy it.

"It may well be the whole thing is due to contamination, but it bugs me everytime I think about it in our own lab," he said a few years later. "Of course, no one is immune . . . yet how it got there and how it could have happened is very mysterious to me."

This was not Relda Cailleau pounding a fist on the table. As he reminisced, he was relaxed, in control, speaking with the quiet intensity of a deep thinker.

"If it *is* due to contamination, then I ask myself, 'Why hasn't it been observed for other cells that are not HeLa? Why don't other cells crisscross? Why only HeLa?'"

These are the same questions Nelson-Rees had been asked a hundred times by scientists hoping to elude the stigma of a HeLa mix-up. So many researchers were under the impression that only HeLa preempted its fellow cell lines. Their argument went: "It just doesn't make sense that no other cell line would misbehave this way, and therefore it must not be happening."

But other cell lines do trespass on their neighbors. All the time. Besides his HeLa studies, Nelson-Rees had published reports listing dozens of nonHeLa mix-ups he had uncovered. Rat cells had overtaken a human breast culture, human cells had infiltrated a gibbon line, hamster cells were in where the marmosets were supposed to be, and dog cells were running wild among the minks. Sure, as an individual contaminator, HeLa had no equal. By no means, however, was it the only cell line that sneaked in where it didn't belong.

But Salk had another theory to offer: Isn't it possible that all cells in long-term culture take on certain common characteristics? We may think of these as HeLa's traits, but maybe they are the universal traits toward which all cells drift. If human beings emerged from monkeys and apes, picking up new traits as they evolved, why couldn't the same thing happen to cells in culture?

"Perhaps something fundamental occurs that expresses genes of evolutionary significance. This would not be a trivial observation," he said. He talked of silent genes gaining voice, "just an intuitive notion," you understand, "still in the form of an hypothesis." And yet maybe the apparent proliferation of HeLa cells is something far more significant. Maybe there's an alternative interpretation to these observations, something monumental, said Salk—like Fleming's recognition of what the penicillin mold was doing to his staphylococcus bacteria!

This was nothing more than a warmed-over version of the argument Stan Gartler got when he unveiled the very first HeLa list in 1966. Wasn't it possible that some cells simply changed their G6PD enzyme from type B to A? Gartler and a Canadian researcher named Nellie Auersperg answered that question by growing one human cell line for three full years, monitoring three of its enzymes closely, and finding no such "evolution." No change even after they subjected the cultures to such triggers of evolutionary change as hormones, bacteria, viruses, and X rays. Cyril Stulberg and Ward Peterson at the Child Research Center in Detroit performed a related experiment a few years later, following the progress of a throat cancer line for two years with no sign of a change in its G6PD.

Similarly, in more than five years spent carefully examining human cell lines, Nelson-Rees never once saw a culture gradually develop HeLa marker chromosomes as if it were evolving into some universal HeLalike beast. When a HeLa takeover happened, it was always just as you'd expect in a cellular coup, quick and complete. Suddenly a culture had HeLa chromosomes, not just one but a whole family of markers, and they were accompanied by HeLa enzymes and other genetic markers that hadn't been there before; an across-the-board changeover, scarcely the kind of gradual metamorphosis that evolution would bring about.

Despite the evidence against it, this evolutionary theory of a "universal drift toward HeLa" had a loyal following.

"I don't doubt for a moment what Nelson-Rees observed in the monkey cells, but it's all a question of interpretation," said Salk. "I still don't want to throw out the notion that it may be due to evolutionary change.

"I'm trying to extract meaning out of these observations. Is this an example of evolution in the lab or is it contamination?"

A thoughtful pause.

"I'll entertain all the possible explanations until we know. I'm enough of a scientist to entertain both explanations."

14

The Little Dutch Boy

The wind blowing through Washington and into the National Cancer Institute suddenly shifted direction in the late 1970s. Part of it was a change in the political climate. Frank Rauscher resigned as director of the institute, and Jimmy Carter installed Arthur Upton as his replacement. Former director of the Brookhaven National Laboratory, Upton had long been interested in the health hazards of radiation. So it was only natural that under his guidance, the institute would become more concerned with the so-called environmental causes of cancer than it was under Nixon's man Rauscher, who was so thoroughly a virologist he even had a virus named after him.

But there was more going on than a simple changing of the chiefs.

By the late '70s, the quest for the viral cause of cancer simply didn't have much to show. There were a lot of amazing finds that stretched the bounds of basic biology and were leading to the genetic engineering revolution of the '80s. But there had been little progress in turning up *the* cause and *the* cure that everyone set out to find when the National Cancer Act was passed in 1971. It was now clear that cancer was a complex family of diseases with many possible causes. The unsung scientists who had been studying nonviral agents figured it was about time they were handed a larger hunk of the pie.

Meanwhile Americans were beginning to learn just how poisonous a land they lived in. In 1976, the institute published terrifying color-coded maps showing that specific cancers were native to certain parts of the country. The Northeast specialized in lung cancer, the Carolinas offered nose and throat cancer, and much of the Midwest and Texas featured leukemia. Suddenly that chemical stench along the northern end of the New Jersey Turnpike took on a horrible new significance, as did the haze over the refineries of the Gulf Coast and, later, those buried drums that turned up in the quiet community of Love Canal, New York. In the public mind, cancer was no longer the work of some mysterious microscopic bug, but something from the "environment," which is to say from the neighborhood—something you could actually see, smell, and taste—and that made it all the more terrifying.

In late 1978 the Carter administration pushed through a new law requiring federal researchers to test the carcinogenic powers of chemicals and low-level radiation. It called for lists to be published every year describing known or suspected cancer agents. And it mandated that the institute expand its study of cancer prevention, including such strategies as proper nutrition and limits on the spread of hazardous industrial materials and environmental pollutants. But because the institute's budget was to grow only slightly, something had to give to make way for the new priorities. And the most vulnerable part of the program was the viral cancer effort.

The resulting slowdown in viral cancer research could easily have threatened the Oakland lab, since cell lines had always been the raw material for viral work. Strangely enough, there were no signs of trouble during the late '70s. Oh, every now and then Jim Duff or some other visiting official would ask Nelson-Rees couldn't the Oakland operation be done on a smaller scale? Or couldn't they move it to Bethesda to be consolidated with some of the institute's other laboratories? But Nelson-Rees, convinced that the lab's prowess stemmed directly from its lavish endowment—in good staff, in quality equipment, in plenty of time to be painstakingly careful—would always argue loudly against it. So loudly, in fact, that even through a closed office door, staffers could hear him: "Just leave us alone so that we can do our jobs properly and well!" Nelson-Rees knew too that any operation ensconced in Bethesda would have none of the freedom that made Oakland so effective. He never raised this point, though. Mostly he ranted about how perfection requires resources, and they always dropped it.

No, the changing winds had not yet reached Oakland. In fact the turn of the decade was looking like boom time for Nelson-Rees's crusade. He and Stevenson finally convinced the editor of the Tissue Culture Association's journal, *In Vitro*, to require a full accounting of cell lines used in published studies. Now an author had to specify the supplier of the cells he studied and the various tests he used to confirm the presumed characteristics of the cells, including species, sex, race, and age of the donor. If he performed no such tests, he had to own up to that in the report. Nelson-Rees and Stevenson considered it a small but significant victory.

What's more, Stevenson left the institute's Frederick Cancer Center in 1979 to take over as director of the American Type Culture Collection. Ever since the early '60s, when he was at the institute trying to make the collection the nation's central cell repository, he felt it hadn't lived up to its first-class potential. As late as 1979 the place was still passing out some of the HeLa-contaminated cultures of the '60s under their original names and without clearly-worded warnings. The collection also had a bountiful supply of genuine cell lines that most researchers didn't seem to know about. As its new director, Stevenson aimed to clean the place up and then promote the hell out of it. With him at the collection and Nelson-Rees at Oakland, two of the world's most valuable biological stockpiles were being cared for by the two most devoted brothers of the Order of the Unspoiled Cell.

As it happened, the turn of the decade was an excellent season for HeLa hunting too. Nelson-Rees, Flandermeyer, and a recently hired technician named David Daniels uncovered HeLa cells masquerading as a liver culture in West Germany. They found HeLa lurking in three supposedly different cell lines from China, which they dubbed, "the gang of three." But of all the investigations that Nelson-Rees and the troops launched in this boom time, the case involving the cells of the little Dutch boy was the most impressive.

In October 1979 researchers at Penn State University published a report in *Science* supporting a controversial claim that tiny amounts of radiation can kill cells, cause genetic damage, and trigger cancer. They exposed human kidney cells known as T-1 to various levels of gamma radiation and found that some were killed even by extremely low doses. Nelson-Rees had never heard of the T-1 cell line, and he was puzzled by a statement in the report suggesting the culture was more than twenty years old. If, as the report implied, these cells were normal— that is, noncancerous—they would probably have gone through their allotted fifty to sixty doublings and died off long ago.

When he called Paul Todd, one of the authors, and identified himself, Todd said, "I had a feeling I would be hearing from you one of these days." Nelson-Rees could scarcely believe it when Todd explained that three years earlier he and his co-workers tried to submit their cell line to the American Type Culture Collection but were turned down because K. S. Lavappa, the collection's chromosome expert, found HeLa markers in it.

"Well then why did you go ahead and describe these as T-1 cells in *Science*?" asked Nelson-Rees. "Why didn't you state that they were HeLa?"

Because when they had tried to mention it in previous reports on the cell line, said Todd, journal reviewers asked that the explanation be deleted. Reviewers for *Radiation Oncology*, for instance, called the information "cell culture folklore" and said it was out of place in their journal. Besides, said Todd, the culture was only suspected of being contaminated; Lavappa's evidence was skimpy. To Todd, these T-1 cells didn't look anything like the HeLa cells he had seen.

Nelson-Rees wanted to judge for himself. He asked Todd to send him a sample of the kidney cells, then orchestrated an analysis so thorough it would be impossible to wave off with a claim of "too skimpy." He set Flandermeyer and Daniels to work on the chromosomes. He enlisted the help of Stephen O'Brien at the institute to test eight key enzymes instead of the usual two or three. And he got a colleague at the Scripps Research Institute in La Jolla to check nine HLA antigens.

While the testing was underway Nelson-Rees pieced together T-1's history. From a 1957 paper reporting the first culturing of the cells, he learned they had come from an eight-year-old Dutch boy who was operated on for kidney stones. A scientist named J. van der Veen put the tissue in a roller tube, where the cells grew for twelve days, then stalled. For months he transferred bits of the culture to new vessels, but none would grow. Then one of the bits suddenly blossomed. The cells spread over the entire tube in seven days and, carved up and placed into additional tubes, kept growing rapidly to become van der Veen's T-1 line.

The report read like such a classic HeLa takeover—van der Veen mentioned he also had been working with HeLa at the time—that Nelson-Rees had no doubt how the test results would turn out. By February 1980 Daniels and Flandermeyer had found four banded HeLa markers in some of the clearest, most convincing mugshots the lab had ever produced; they also reported the lack of any Y chromosomes, despite the fact that the donor was a boy. All nine HLA antigens tested in these "kidney cells" matched those of HeLa, as did all eight enzyme variants; the odds of these biochemical fingerprints being identical purely by chance were less than one in a million.

It was an airtight case against the Penn State culture. Having dug through the literature, though, Nelson-Rees realized that University Park, Pennsylvania, was only one of many stops the T-1 cell line had made. Over twenty years various strains of T-1 had become the favorites of radiation health researchers all around the world. To Nelson-Rees it looked as if a T-1 mafia had quietly gained control of the entire field. Much of what was known about the harm inflicted on human cells by radiation, and much of what influenced safety standards and guidelines to radiation exposure, was apparently based on experiments with T-1. He decided the case could not be closed until he had rounded up the entire gang.

The Penn State cells were ancestors of a sample of T-1 that Todd had obtained from researchers at Lawrence Berkeley Laboratory in California, who had also sent some to colleagues at the Los Alamos Scientific Laboratory in New Mexico. The Berkeley scientists got theirs from a Dutch researcher named G.W. Barendsen, who got his from van der Veen himself. Nelson-Rees retraced the trail, asking for samples from every lab along the way; he even got van der Veen to send a culture from a frozen stock of very early vintage. In August of 1980, *Science* published his reconstruction of the whole affair. It was an investigative tour de force. Five strains of T-1 cells from laboratories as far apart as Berkeley and Utrecht showed precisely the same incriminating characteristics of HeLa.

Science also ran a news analysis like the one that accompanied Nelson-Rees's first hit list. Titled "The Case of the Unmentioned Malignancy," the article made hay of the Penn State researchers' decision not to report the HeLa contamination of their cell line. Nelson-Rees tried to be tactful when asked for his comment. He noted the attempts by Todd's group to present the information in other, earlier reports. "I don't think that they swallowed the whistle," he was quoted as saying. "But they certainly didn't blow it." A couple of newspapers and magazines played the incident like a minor scandal.

The question of what HeLa's latest charade really meant for radiation research and existing standards, unfortunately, was never settled. Todd and associates claimed that even if T-1 were a HeLa culture, it wouldn't change their results. There was no evidence that a cancer cell would respond to radiation any differently from a normal cell, they said. One biophysicist at the Argonne National Laboratory, however, *had* observed that tumor cells survive some types of radiation significantly better than normal cells. She wrote an angry letter to *Science* complaining about the Penn State study. Bob Stevenson joined the fray too, writing to Todd's group that it was "pretty shoddy" to use HeLa-contaminated cells in an experiment without revealing it. He was so angry, in fact, he also wrote to the institute's grant officers who were funding the Penn State team to make sure they knew about it.

All in all, it was a highly visible demonstration of the Oakland team's powers of investigation and prosecution. In celebration, Daniels and a few of the other lab workers presented Nelson-Rees with a small gardening tool labeled "official muck rake."

Just about that time, an ominous bunch of letters went out from the institute's biological carcinogenesis branch, the direct descendant of the old viral cancer office. In light of "possible budget reductions," the letters said, the institute was polling people who recently had obtained cells from Nelson-Rees or had asked his lab to identify unknown cultures. How heavily did these people depend on the cells and the expertise in Oakland, the letters asked, and what impact would it have on their work if they had to pay a commercial laboratory for these services, which the institute had been supplying to them for free?

Nelson-Rees got a letter too. It directed him to describe the lab's activities and explain why they were necessary.

Were they sincere? Did they actually have to be told why it was necessary to safeguard the most widely used experimental material in cancer research? Could it be, Nelson-Rees wondered, that they were unacquainted with one of the basic principles of the scientific method: to know what the hell one is working with? And was Walter Nelson-Rees, the keeper of the cells, actually being ordered to *account* for himself? It was an outrage.

He considered sending out his own letter to researchers asking how heavily they depended on the institute's biological carcinogenesis branch and what would be the impact of phasing out a few of the bureaucrats there. But then it occurred to him that maybe this wasn't simply another argument he would win by shouting down Jim Duff. Maybe this time they were serious.

In fact, for the first time since America declared war on cancer, the institute's total budget was about to shrink: from $958 million in 1980 to $947 million for the upcoming year. The funding for viral research and related activities, already on the slide, would be cut back even more.

The shifting winds from Washington had finally reached Oakland. Shortly before Christmas 1980, the bureaucrats notified Nelson-Rees that his $600,000 annual budget would be trimmed by 20 percent. He would have to fire six of his fifteen people.

15

Battle Fatigue

In February of 1981, Bob Flandermeyer attended a conference in San Francisco on genetic engineering, which, unlike most scientific conferences Flandermeyer had been to, was crawling with corporate researchers. The manipulation of genes was becoming big business.

Flandermeyer went to a talk about interferon, a natural protein that showed some ability to fight off viruses and control the spread of certain cancers. Recently featured on the cover of *Time*, interferon was now being synthesized by many of the new firms, whose directors hoped to cash in on its wonder-drug potential. Flandermeyer heard a fellow from Genentech, one of the giants in the field, describe how they tested the potency of their interferon. They treated several cultures of cells with different doses of the chemical, then rated how well those cells withstood assaults by hostile viruses. The cell line routinely used for the test, said the man from Genentech, was a culture of normal human amnion tissue called WISH.

Flandermeyer thought WISH sounded familiar, but before saying anything he telephoned the lab.

"WISH, of course!" bellowed Nelson-Rees. "Wistar Institute Susan Hayflick. That was the amniotic sac in which Leonard Hayflick's daughter was delivered. He made a big joke about it when Stan Gartler claimed it was HeLa in 1966."

Flandermeyer said he'd better go tell the genetic engineer.

Over the next few weeks, Nelson-Rees thought a lot about that simple phone conversation. It occurred to him that the current boom in bioengineering was very different from the biomedical boom in the '70s. The motivation then was "The War." The glorious dream of conquering cancer had pumped billions of dollars into research and attracted hundreds of eager recruits. Now the motivation was profit. These outfits were businesses, and the pressure to be the first with a new product stemmed from commercial competition. Furthermore, proprietary information and corporate secrets were the standard rules by which these new people played. Businessmen don't openly share the details of their manufacturing processes, even if their factories are living cells and bacteria, their foremen molecular biologists.

Not that there was anything wrong with Genentech's using HeLa cells to test interferon. Since the cells were serving strictly as targets for viruses, the potency ratings were probably valid as long as the company always used the same cell line. What bothered Nelson-Rees was that the people at Genentech were calling these cells human amnion. *They didn't know.* They might never have known if Bob Flandermeyer hadn't wandered in and told them. And if they didn't know in this case, they might not know in cases where it really matters.

After all, these researchers were playing around with genes, snipping a few out of one cell and splicing them in among those of another. Well, scientists knew enough to say that genes control everything, but they had barely a clue about which of the thousands of individual genes directed this function or determined that characteristic.

Cancer researchers, for instance, had just lately found out about the oncogenes, a mysterious bunch of genetic elements that usually sit dormant upon the human chromosome but that may under certain conditions in certain cells "turn on" and trigger cancer. Which only meant to Nelson-Rees that as these bioengineers are cutting and pasting up genes to create hormones and drugs and other things intended to go into human bodies, they'd better be damned sure they know what cells they're finagling.

That one short phone call from Flandermeyer made Nelson-Rees realize this was probably one of the worst times to abandon the crusade, and yet that's exactly what he was considering doing. It had been frustrating for him these last few months: trying to run the cell culture lab at less than full strength and at the same time keep the crusade going. Everyone was putting in extra hours. Still he knew the cutbacks had only begun, and any added load would be too much to handle if they were to keep up the quality of work that the lab—that he—had always been known for. David Daniels said the place was starting to take on the look of "a ma-and-pa operation." That phrase stuck with Nelson-Rees, grating on his perfectionism like a burr in his boot.

The funding cutback was really the last of a series of frustrations. Although he rarely showed it and never discussed it even with close friends, Nelson-Rees was weary of the battle. He was in a business that generated enemies, not allies. Even the Tissue Culture Association, the scientific group that ought to have been cheerleading him on, only tolerated him after the blow-up in Denver several years earlier. The latest order from the institute to defend the lab's existence in spite of what he felt was a glorious record reminded him once again that not everyone thought his contribution was so vital.

"I don't think Walter felt anyone fully appreciated the value of the work he had done," Jim Duff once observed. "You know, they don't award Nobel Prizes for finding out that things are wrong."

So in March Nelson-Rees called the staff together and gave a subdued little speech about the cutbacks, present and future, saying in short that you can't fight a crusade without funding. Difficult as it would be, he said, he would be quitting toward the end of the year. Not just quitting the lab and moving on, but leaving research entirely. He planned to devote his time to the small art gallery he and a friend had been operating out of his home the past few years.

Quitting, just like that.

Nelson-Rees figured a scaled-down cell culture lab would continue at Oakland. But after his resignation the bureaucrats announced that the facility would be shut down completely by the end of the next year. It looked as if he had played right into their hands. They said they would arrange to have the 2,000 cell cultures transferred to the freezers of the American Type Culture Collection, although they could provide no funds for the maintenance and distribution of the cells. As for the Oakland lab's service of checking the identity of cultures, they weren't setting up any substitute for that. Ward Peterson in Detroit could do the enzyme testing he had always done as well as some of the simpler chromosomal checks. But researchers would have to pay for the analysis.

"I'll have to think very carefully now before sending some culture out to be checked," said one candid scientist when asked about the loss of the Oakland lab. "When it was all free, we sent things out all the time. It was a tremendous help. But now, at two or three hundred dollars a shot . . . ."

Even more than the loss of the service, though, some mourned the loss of the crusader himself. Peterson wrote to Nelson-Rees, "As the most articulate amongst a small group that has promulgated the rules, your role as the point man will greatly be missed." That was the real worry of the few cancer researchers and cell biologists who had appreciated Nelson-Rees. Who was to carry on? Peterson was concerned about cell line screw-ups, but he was no zealot. Stevenson would continue his missionary work, of course, in a tactful and politically sensitive way. But challenging colleagues at public meetings, naming names in journals, making an occasional scene—that just wasn't their style.

There were a few protests to the institute. One scientist called Nelson-Rees "a national resource," and estimated that by spotting suspicious cell lines he had probably prevented the waste of tens of millions, if not hundreds of millions of research dollars. The institute was unimpressed. The feeling, at least among certain officials, was that these worries were held over from the old days, the unenlightened days when cell culture needed to be actively policed.

"Nelson-Rees may have been exactly the right person for that time," explained one bureaucrat. "A more subtle and conciliatory style might not have spread knowledge of the extent of HeLa contamination throughout the world back then. Today, however, I would hope that there is sufficient awareness and sufficient expertise that any of these potential things would be avoided or caught quickly."

With Nelson-Rees off the scene, that was about all they could do: hope that nothing went wrong.

16

Legacy

In a cramped and cluttered office at Montefiore Hospital in the Bronx, there stands a kind of monument to Nelson-Rees and his crusade. Tacked onto the door frame is a slightly yellowed copy of his third and final hit list, published in April 1981, just a few weeks after he announced he was quitting.

This is the office of Fritz Herz, chief of the tissue culture section in the hospital's pathology department. Five times a year, Herz holds a seminar in the art of culturing cells. He starts with a demonstration of the basics, reminding his students never to feed different cells from the same pipette—"Look," he tells them, "I'm throwing the pipette away"—and ends with the story about Henrietta Lacks, Walter Nelson-Rees, and the list.

Things are not always what they seem, Herz explains, pointing to the three full pages that feature such oldies as WISH and Relda Cailleau's MAC-21 plus twenty-two new HeLa-contaminated cultures uncovered by Nelson-Rees, Flandermeyer, and Daniels in the last couple of years they worked together. Ninety cell lines in all. Bob Stevenson guessed that was about a third of the more popular cell lines used in cancer and related research: one out of every three cell lines was an impostor.

"It's very important that you know what your cells are, that you know they are not HeLa," Herz tells the residents, the neuropathologists, and the other young scientists who come through his lab. "If you're sloppy and you don't care and you're not committed to perfection, you'll soon find your own work printed on a list like this."

That was how Nelson-Rees left his mark. A few loyal fans, a couple of disciples who preached in their own small parishes. He had untangled much of the morass created by Henrietta Lacks's runaway cells, but once he left his post, there was no guarantee that things would stay untangled. The bureaucrats were hoping that the collective consciousness of medical research had been propped up to where it no longer needed to be nagged about sloppy technique and wasteful mix-ups. But if ever there was a persuasive argument against that kind of thinking, this list was it. Just because Nelson-Rees had got out of the business, its ninety entries seemed to say, that didn't mean the ghost of Henrietta Lacks was retiring.

Which proved to be absolutely correct.

Microbiological Associates, for example, the company that created the HeLa-contaminated culture called MA160 thirteen years before the publication of this last list, had changed its name to M.A. Bioproducts but made no change in its catalogue description of MA160. Although Nelson-Rees and half-a-dozen independent scientists had concluded by the mid-'70s that the culture was nothing but cervical cancer cells from a black woman, the company continued selling it through 1981 as "prostate, benign, human adult."

In the fall of that same year, as Nelson-Rees was cleaning out his file cabinets in Oakland, a new study on prostate cancer was published. It was based on experiments with MA160 and its identical, HeLa-contaminated twin EB33, the culture Nelson-Rees had condemned in

1976. The following year, there appeared another study based in part on these two adulterated cultures.

In January 1983 a team of Dutch, Finnish, and American researchers announced their discovery that the U cell line, presumed to be a culture of amnion cells taken from a Dutch woman in 1957, was actually HeLa.

And in the spring, Nelson-Rees spoke with a young San Diego researcher named Mark Bogart, who had recently checked thirty different cell lines from various scientists and found that half were not what these scientists had thought. Bogart's methods were unsophisticated compared to the techniques Nelson-Rees had used. Like the early tissue culturists in the 1950s, he could only tell species apart. He knew that two mouse lines had been jumped by a culture of human cells, for example, but he couldn't say which human culture that was. He knew that ten of the supposedly human cultures he analyzed were indeed human, but he had no idea whether they held the *right* human cells or—who knows?—perhaps the cells of a certain lady from Baltimore.

17

Epilogue

Walter Nelson-Rees is no longer chasing Henrietta Lacks, but he keeps in touch with a few colleagues and peruses the journals regularly enough to know that she still haunts the biomedical labs. And sitting in his home in Oakland, next to a wall filled with impressionist paintings, he is happy to deliver a private sermon when asked why she endures.

There's more to the problem than a tenacious and hardy cell culture, he says. HeLa cells persist because they have always been helped along by a certain human element in science, an element connected to emotions, egos, a reluctance to admit mistakes, and many of the other things he used to lecture about.

"It's all human—an unwillingness to throw away hours and hours of what was thought to be good research, worries about jeopardizing another grant that's being applied for, the hurrying to come out with a paper first. And it isn't limited to biology and cancer research. Scientists in many endeavors all make mistakes, and they all have the same problems."

Wade Parks, the virologist who in 1972 exposed the famous Russian cells as worthless carriers of a monkey virus, once said something similar. In every field, he observed, this human factor encourages "HeLas."

"A 'HeLa,'" Parks explained, "is a scientific claim that sucks people into a line of work for a while, a line that is later refuted or shown to be a waste of time. It's a type of error in science that occurs fairly often. And it will continue to exist."

Acknowledgments

The writing of this book depended upon the help of scores of people, each of whom spent hours with me, recalling events, emotions, even conversations that took place as long ago as twenty years. For their willingness to relive those times and to be candid about them, I am very grateful. My thanks also to Robert Stevenson of the American Type Culture Collection for allowing me access to the files of the former Berkeley Cell Culture Laboratory, of which he is now the custodian. I am grateful, too, to the people at *Science 85* magazine for allowing me a brief leave of absence to get started on this book and for putting up with me after I returned to work and continued writing it in my "spare time." Many thanks to Jack Godler, who cajoled me into undertaking this project in the first place, and to Susan Zeckendorf, who shared her knowledge of publishing and her great enthusiasm.

My special gratitude goes to two people who served as unofficial editors and tireless cheerleaders: Susan West and Eric Schrier. Their detailed critiques and suggestions greatly improved every draft chapter. And without their genuine support, this book might never have been completed. Finally, to Susan, who while editing and cheerleading also offered love and patience, my greatest thanks.

Sources

The following references offer some historical perspective to the events described, as well as greater technical depth for those who are interested. It is by no means comprehensive.

Chapter 1. Special Delivery

"Isolation and Culture of a Virus from Human Cancer Tissue." *Science News*, 10 July 1971, p. 21.

"Race for Human Cancer Virus: Odds against Houston Team Lengthen." *Science*, 24 September 1971, p. 1220.

"USA and USSR Communication in Cancer Research," by Joseph F. Saunders, *Biosciences Communications*, 2, volume 2, (1976) p. 98.

Chapter 2. The Seed That Took

"Some Aspects of the Constitution and Behavior of Normal and Malignant Cells Maintained in Continuous Culture," by George O. Gey, *The Harvey Lectures: Series L (1954-1955)*, Academic Press Inc., New York, N.Y., p. 154.

"George Otto Gey: The HeLa Cell and a Reappraisal of its Origin," by Howard Jones Jr., Victor A. Mckusick, Peter S. Harper, and Kuang-Dong Wuu, *Obstetrics and Gynecology*, December 1971, p. 945.

"History of Tissue Culture at Johns Hopkins," by Frederick B. Bang, *Bulletin of the History of Medicine*, volume 51, 1977, p. 516.

Chapter 3. HeLagram

"Studies on the Propagation in Vitro of Poliomyelitus Viruses," by William F. Scherer, Jerome Syverton, and George O. Gey, *Journal of Experimental Medicine*, volume 97, 1953, p. 695.

"Continuous subcultivation of epithelial-like cells from normal human tissues," by R.S. Chang, *Proceedings of the Society for Experimental Biology and Medicine*, volume 87, 1954, p. 440.

"The origin of altered cell lines from mouse, monkey, and man, as indicated by chromosome and transplantation studies," by K.H. Rothfels, A.A. Axelrad, L. Siminovitch, E.A. McCulloch, and R.C. Parker, *Proceedings of the Third Canadian Cancer Conference*, 1959, p. 189.

"The establishment of a line (WISH) of human amnion cells in continuous cultivation," by Leonard Hayflick, *Experimental Cell Research*, volume 23, 1961, p. 14.

"Results of tests for the species of origin of cell lines by means of the mixed agglutination reaction," by D. Franks, B.W. Gurner, R.R.A. Coombs, and R. Stevenson, *Experimental Cell Research*, volume 28, 1962, p. 608.

"Collection, Preservation, Characterization, and Distribution of Cell Cultures," by Robert E. Stevenson, *Proceedings of the Symposium on the Characterization and Uses of Human Diploid Cell Strains*, Opatija, Yugoslavia, 1963, p. 417.

"Genetic Markers as Tracers in Cell Culture," by Stanley M. Gartler, *National Cancer Institute Monograph No. 26: Second Decennial Review Conference on Cell Tissue and Organ Culture*, 1967, p. 167.

"Apparent HeLa Cell Contamination of Human Heteroploid Cell Lines," by Stanley M. Gartler, *Nature*, February 24, 1968, p. 750.

"Cell Culture Collection Committee in the United States," by Robert E. Stevenson, *Cancer Cells in Culture*, edited by H. Katsuta, University of Tokyo Press, 1968, p. 385.

"Extrinsic Cell Contamination of Tissue Cultures," by Cyril S. Stulberg, *Contamination in Tissue Culture*, edited by Jorgen Fogh, Academic Press, New York and London, 1973, p. 1.

Chapter 4. Out of Thin Air

"Oncogenesis in Vitro," by Leonard Hayflick, *National Cancer Institute Monograph No. 26: Second Decennial Review Conference on Cell Tissue and Organ Culture*, 1967, p. 355.

"Nutritional Needs of Mammalian Cells in Tissue Culture," by Harry Eagle, *Science*, September 16, 1955, p. 501.

"Common Antigens in Tissue Culture Cell Lines," by Lewis L. Coriell, Milton G. Tall, and Helen Gaskill, *Science*, July 25, 1958, p. 198.

"Detection and Elimination of Contaminating Organisms," by L. Coriell, *National Cancer Institute Monograph No. 7*, April 1962, p. 33.

"Glucose-6-phosphate Dehydrogenase Isoenzymes in Human Cell Cultures Determined by Sucrose-Agar Gel and Cellulose Acetate Zymograms," by W.D. Peterson Jr., C.S. Stulberg, N.K. Swanborg, and A.R. Robinson, *Proceedings of the Society of Experimental Biology and Medicine*, volume 128, 1968, p. 772.

"The Animal Cell Culture Collection," by C.S. Stulberg, L.L. Coriell, A.J. Kniazeff, and J.E. Shannon, *In Vitro*, volume 5, 1970, p. 1.

"Oncornavirus-like particles in HeLa Cells," by H. Bauer, J.H. Daams, K.F. Watson, K. Molling, H. Gelderblom, and W. Schafer, *International Journal of Cancer*, volume 13, 1974, p. 254.

"A Quest for the Mechanism of 'Spontaneous' Malignant Transformation in Culture with Associated Advances in Culture Technology," by Katherine K. Sanford and Virginia J. Evans, *Journal of the National Cancer Institute*, June 1982, p. 895.

Chapter 5. In The Purple Palace

"Isolation of Oncornaviruses from Continuous Human Cell Cultures," by Victor M. Zhdanov, Valentine D. Soloviev, Tagir A. Beketemirov, Konstantin V. Ilyin, Albert F. Bykovsky, Nikolai P. Mazurenko, Iosif S. Irlin, and Felix I. Yershov, *Intervirology*, volume 1, 1973, p. 19.

"Mason-Pfizer virus characterization: A similar virus in a human amniotic cell line," by W.P. Parks, R.V. Gilden, A.F. Bykovsky et al., *Journal of Virology*, volume 12, 1973, p. 1540.

"HeLa-Like Marker Chromosomes and Type-A Variant Glucose-6-phosphate Dehydrogenase Isoenzyme in Human Cell Cultures Producing Mason-Pfizer Monkey Virus-Like Particles," by W.A. Nelson-Rees, V.M. Zhdanov, P.K. Hawthorne, and R.R. Flandermeyer, *Journal of the National Cancer Institute*, September 1974, p. 751.

Chapter 7. Mug Shots

"Quinacrine fluorescent karyotypes of human diploid and heteroploid lines," by O.J. Miller, D.A. Miller, P.W. Allderdice, V.G. Dev, and M.S. Grewal, *Cytogenetics*, volume 10, 1971, p. 338.

"Isolation of a Continuous Epithelioid Cell Line, HBT-3, from a Human Breast Carcinoma," by Robert H. Bassin, Ernest J. Plata, Brenda I. Gerwin, Carl F. Mattern, Daniel K. Haapala, and Elizabeth W. Chu, *Proceedings of the Society for Experimental Biology and Medicine*, November 1972, p. 673.

"An Established Cell Line (HBT-39) From Human Breast Carcinoma," by Ernest J. Plata, Tadao Aoki, Diane D. Robertson, Elizabeth W. Chu, and Brenda I. Gerwin, *Journal of the National Cancer Institute*, April 1973, p. 849.

"Banded Marker Chromosomes as Indicators of Intraspecies Cellular Contamination," by Walter A. Nelson-Rees, Robert R. Flandermeyer, and Paula K. Hawthorne, *Science*, June 7, 1974, p. 1093.

"HeLa Cells: Contaminating Cultures around the World." by Barbara J. Culliton, *Science*, June 7, 1974, p. 1059.

"Human Breast Tumor Cell Lines: Identity Evaluation by Ultrastructure," by Gertrude C. Buehring and Adeline J. Hackett, *Journal of the National Cancer Institute*, September 1974, p. 621.

Chapter 8. Spreading the Word

"The establishment of a Cell Strain (MAC-21) from a Mucoid Adenocarcinoma of the Human Lung," by Relda Cailleau, *Cancer Research*, July 1960, p. 837.

"Two Cell Lines (SH-2 and SH-3) Derived from Human Breast Cancer," by G. Seman, S.J. Hunter, and L. Dmochowski, *Proceedings of the American Association for Cancer Research*, volume 16, March 1975, p. 59.

Chapter 9. Damage Report

"Detection and Isolation of a New DNA Polymerase from Human Breast Tumor Cell Line HBT-3 by (dT) 12-18-Cellulose Chromatography," by Brenda I. Gerwin and Robert H. Bassin, *Proceedings of the National Academy of Sciences*, August 1973, p. 2453.

"Present Status of MA160 Cell Line, Prostatic Epithelieum or HeLa Cells?" by Mukta M. Webber, Paul K. Horan, and Thomas R. Bouldin, *Investigative Urology*, volume 14, 1977, p. 335.

Chapter 10. Provenance

"Spontaneous in vitro Neoplastic Transformation of Adult Human Prostatic Epithelium," by Elwin E. Fraley, Sidney Ecker, and Monroe M. Vincent, *Science*, October 30, 1970, p. 541.

"Human Prostatic Carcinoma in Cell Culture: Preliminary Report on the Development and Characterization of an Epithelial Cell Line (EB 33)," by K. Okada and F.H. Schroeder, *Urological Research*, volume 2, 1974, p. 111.

"Human Breast Cancer Cells in Continuous Cultivation Used to Determine Chemotherapy Sensitivity of the Patient from Whom They Are Derived," by Roland A. Pattillo and A.C.F. Ruckert, *Proceedings of the American Association for Cancer Research*, volume 16, March 1975, p. 145.

"HeLa Cultures Defined," by Walter A. Nelson-Rees and Robert R. Flandermeyer, *Science*, January 9, 1976, p. 96.

"Examination of ATCC stocks for HeLa marker chromosomes in human cell lines," by K.S. Lavappa, M.L. Macy, and J.E. Shannon, *Nature*, January 22, 1976, p. 211.

"HeLa takes over," by Sandy Grimwade, *Nature*, January 22, 1976, p. 172.

"Cell Cultures: Confused and Contaminated," *Science News*, volume 109, 1976, p. 36.

"Characterization of an established cell line (SH 3) derived from pleural effusion of patient with breast cancer," by G. Seman, S.J. Hunter, R.C. Miller, and L. Dmochowski, *Cancer*, volume 37, 1976, p. 1814.

Chapter 11. Another Run-in With Relda

"Human Lung Organ-Specific Antigens on Normal Lung, Lung Tumors, and a Lung Tumor Cell Line," by Richard Akeson, *Journal of the National Cancer Institute*, April 1977, p. 863.

"Lung Organ-Specific Antigens on Cells with HeLa Marker Chromosomes," letters by Walter A. Nelson-Rees, R.A. Akeson, and Relda Cailleau, *Journal of the National Cancer Institute*, June 1978, p. 1205.

Chapter 12. Showdown

"HeLa cells and their possible contamination of other cell lines: Karyotype studies," by W.K. Heneen, *Hereditas*, volume 82, 1976, p. 217.

"Foreword," by T.C. Hsu, *Mammalian Chromosome Newsletter*, November 4, 1978, p. 103.

"A Human Breast Adenocarcinoma With Chromosome and Isoenzyme Markers Similar to Those of the HeLa Line," by Sen Pathak, Michael J. Siciliano, Relda Cailleau, Charles L. Wiseman, and T.C. Hsu, *Journal of the National Cancer Institute*, February 1979, p. 263.

"A Molecular Approach to the Identification and Individualization of Human and Animal Cells in Culture: Isozyme and Allozyme Genetic Signatures," by Stephen J. O'Brien, John E.

Shannon, and Mitchell H. Gail, *In Vitro*, volume 16, 1980, p. 119.

Chapter 13. Even the Best of Labs

"Some Characteristics of a Continuously Propagating Cell Derived from Monkey Heart Tissue," by Jonas E. Salk and Elsie N. Ward, *Science*, December 27, 1957, p. 1338.

"Immunological Paradoxes: Theoretical Considerations in the Rejection or Retention of Grafts, Tumors, and Normal Tissue," by Jonas Salk, *Annals of the New York Academy of Sciences*, October 14, 1969, p. 365.

"Isozyme stability in human heteroploid cell lines," by N. Auersperg and S.M. Gartler, *Experimental Cell Research*, volume 61, 1970, p. 465.

"A Permanent Heteroploid Human Cell Line with Type B Glucose-phosphate Dehydrogenase," by W.D. Peterson Jr., C.S. Stulberg, and W.F. Simpson, *Proceedings of the Society for Experimental Biology and Medicine*, April 1971, p. 1187.

"Inter- and Intraspecies Contamination of Human Breast Tumor Cell Lines HBC and BrCa5 and Other Cell Cultures, by Walter A. Nelson-Rees and Robert R. Flandermeyer, *Science*, March 25, 1977, p. 1343.

"The Identification and Monitoring of Cell Line Specificity," by Walter A. Nelson-Rees, *The Origin and Natural History of Cell Lines*, edited by Claudio Barigozzi, Alan R. Liss, Inc., New York, 1978, p. 25.

"The Spector of Malignancy and Criteria for Cell Lines as Substrates for Vaccines," by Jonas Salk, *Cell Substrates and Their Use in the Production of Vaccines and Other Biologicals*, edited by John C. Petricciani, Hope E. Hopps, and Paul Chapple, Plenum Press, New York and London, 1979, p. 107.

"Characteristics of HeLa strains: Permanent vs. variable features," by W.A. Nelson-Rees, L. Hunter, G.J. Darlington, and S.J. O'Brien, *Cytogenetics and Cell Genetics*, volume 27, 1980, p. 216.

Chapter 14. The Little Dutch Boy

Atlas of Cancer Mortality for U.S. Counties: 1950 - 1969, by Thomas J. Mason, Frank W. McKay, Robert Hoover, William J. Blot, and Joseph F. Fraumeni Jr., U.S. Department of Health, Education, and Welfare Publication No. (NIH) 75-780.

"Establishment of Two Human Cell Strains from Kidney and Reticulosarcoma of Lung," by J. van der Veen, L. Bots, and A. Mes, Archiv fur die Gesamte Virusforschung, volume 8, 1958, p. 230.

"Comparison of the Effects of Various Cyclotron-Produced Fast Neutrons on the Reproductive Capacity of Cultured Human Kidney (T-1) Cells," by Paul Todd, Joseph Geraci, Paul S. Furcinitti, Randall M. Rossi, Fuminori Mikage, Richard B. Theus, and Carter B. Schroy, International Journal of Radiation Oncology, Biology, Physics, volume 4, 1978, p. 1015.

"The Effects of Caffeine on the Expression of Potentially Lethal and Sublethal Damage in Gamma-Irradiated Cultured Mammalian Cell," by Carter B. Schroy and Paul Todd, Radiation Research, volume 78, 1979, p. 312.

"Inactivation of Human Kidney Cells by High-Energy Monoenergetic Heavy-Ion Beams," by Eleanor A. Blakely, Cornelius A. Tobias, Tracy C.H. Yang, Karen C. Smith, and John T. Lyman, Radiation Research, volume 80, 1979, p. 122.

"Gamma Rays: Further Evidence for Lack of a Threshold Dose for Lethality to Human Cells," by Paul S. Furcinitti and Paul Todd, Science, October 26, 1979, p. 475.

"T-1 Cells Are HeLa and Not of Normal Human Kidney Origin," by Walter A. Nelson-Rees, Robert R. Flandermeyer, and David W. Daniels, Science, August 8, 1980, p. 719.

"Hendrik or Henrietta?" The Economist, November 15, 1980, p. 104.

"The Case of the Unmentioned Malignancy," by William J. Broad, Science, December 12, 1980, p. 1229.

Chapter 16. Legacy

"Cross-Contamination of Cells in Culture," by W.A. Nelson-Rees, D.W. Daniels, and R.R. Flandermeyer, Science, April 24, 1981, p. 446.

"The Surface Character of Separated Prostatic Cells and Cultured Fibroblasts of Prostatic Tissue as Determined by Concanavalin-A Hemadsorption," by K. Oishi, J.C. Romijn, and F.H. Schroeder, The Prostate, volume 2, 1981, p. 11.

"Monoclonal Antibodies to Human Prostate and Bladder

Tumor-associated Antigens," by James J. Starling, Susan M. Sieg, Mary L. Beckett, Paul F. Schellhammer, Leopoldo E. Ladaga, and George L. Wright Jr., *Cancer Research*, August 1982, p. 3084.

"U cells contain contaminants," by William C. Wright, Sara Kaffe, Christian F. Holinka, Kari Cantell, Jacoba G. Kapsenberg, and Kurt Hirschhorn, *Nature*, January 27, 1983, p. 279.

Index

www.ingramcontent.com/pod-product-compliance
Lightning Source LLC
Chambersburg PA
CBHW021559210326
41599CB00010B/514